Pankaj K. Godhaviya

Fluorinating Reagents

Properties, Preparation, Applications And Safety

Preface

The work to be presented in this book entitled "Fluorinating Reagents" is divided into three sections which can be summarised as under. Section-1 deals with the Introduction of organofluorine chemistry, Section-2 covers Types of Fluorinating Reagents. Section-3 describes the different available Fluorinating reagents, in this section we describe Fluorinating reagents chemical, physical properties, Preparation method, it's Application and safety for handling of Fluorinating Reagents. That's why this book will be useful for the synthesis of organofluorine compounds.

'No serious effort in life is totally accomplished by oneself '. This book work is no exception. The completion of this project was aided by the contributions of a number of people. First and foremost we bow before the almighty God for showering his blessing on us and giving us the strength to carry out the present work with utmost dedication and enthusiasm. Without that the present work would not have been possible. We express our sincere thanks to our respected parents, family members and friends for their constant encouragement, inspiration and untiring efforts for completion of this project.

I acknowledge the crucial assistance and support from the CreateSpace. Finally I would like to express my sincere gratitude towards my college at Navin Fluorine International Ltd. (NFIL), Surat. They were helped me in collecting data, express his experience regarding fluorine chemistry, with their helpful support to complete this project without which this task could not have been achieved.

Thanks to all.

Pankaj K. Godhaviya

About The Authors

Pankaj K. Godhaviya

The author received his B.Sc., M.Sc. and Ph.D. degree in Chemistry, at the Saurashtra University at Rajkot. He has participated and presented scholarly research papers in national and international seminars. Three books are authored by him in Internationally reputed publishing house (1) Chalcones and Isoxazoles Scholars Press ISBN 9783639718799, (2) 1,2,4-Oxadiazoles-Scholars Press-ISBN-9783639661033), (3) Studies on 1,3-Thiazole derivatives (Biological Activity, Synthetic Aspects, Reaction Mechanism and Spectral Discussion) and he also published four Research articles in internationally reputed journals. He has published 4 research papers in various reputed Journals. He is currently working as a Senior Research Associate in Research and Development (CRO) Division of Navin Fluorine International Limited, Surat, Gujarat (India). In past he had also worked with Jubilant Chemsys Limited, Noida (U.P.) (Medicinal chemistry division) and with Cadila Pharmaceuticals Limited, Ahmedabad (R & D-Peptide Chemistry Division).

Contents   **Page No.**

Abbreviations

NaOH	Sodium hydroxide
EtOH	Ethyl alcohol
CS_2	Carbon disulfide
NH_3	Ammonia
KOH	Potassium hydroxide
H_2SO_4	Sulfuric acid
$AlCl_3$	Aluminium chloride
BrCN	Cyanogen Bromide
$NaHCO_3$	Sodium Bicarbonate
CDI	Carbonyl Diimidazole
CBr_4	Carbon Tetrabromide
Et_3N	Triethyl Amine
HCl	Hydrochloric Acid
$SOCl_2$	Thionyl Chloride
$POCl_3$	Phosphoryl Trichloride
CCl_4	Carbon Tetrachloride
$BF_3.Et_2O$	Boron Trifluoride Etherate
Ac_2O	Acetic Anhydride
AcOH	Acetic Acid
NaOAc	Sodium Acetate
% w/v	Percentage Weight by Volume
% v/v	Percentage Volume by Volume
$CDCl_3$	Deuteriated Chloroform
DMSO	Dimethyl Sulfoxide
ppm	Part Per Million
DMF	Dimethyl Formamide
THF	Tetrahydrofuran
BuLi	Butyl Lithium
$ZnCl_2$	Zinc Chloride
TFA	Trifluoroacetic Acid
HCOOH	Formic Acid
MF	Molecular Formula
MW	Molecular weight
MHz	Mega Hertz
CH_2Cl_2	Dichloromethane
DCM	Dichloromethane

HF	Hydrofluoric acid
LD	Lethal Dose
KBr	Potassium bromide
DAST	Diethylaminosulfur trifluoride
MeCN	Acetonitrile
NH_4F	Ammonium fluoride
NaH	Sodium hydride
Rt	Room temperature
MeOH	Methyl alcohol

Chapter-1

Organofluorine chemistry

1.1. Organofluorine chemistry:

Organofluorine chemistry describes the chemistry of organofluorine compounds, organic compounds that contain the carbon–fluorine bond. Organofluorine compounds find diverse applications ranging from oil and water repellents to pharmaceuticals, refrigerants and reagents in catalysis. In addition to these applications, some organofluorine compounds are pollutants because of contributions to ozone depletion, global warming, bioaccumulation, and toxicity. The area of organofluorine chemistry often requires special techniques associated with the handling of fluorinating agents.

The introduction of fluorine into bioactive molecules often results in significant changes in their chemical, physical, and biological properties. Fluorine uniquely affects the property of organic molecules due to the fluorine atom's blocking effect in metabolic transformations and mimicking of enzyme substrates, and then increases the molecular lipophilicity to enhance bioavailability. Approximately 30% of all agrochemicals and 20% of all pharmaceuticals contain fluorine.1) HMG-CoA reductase inhibitor fluorostatin (atorvastatin), antibiotic fluoroquinolone (levofloxacin), and antitumor fluoronucleoside (tegafur) are successful examples of the introduction of fluorine into organic molecules. Therefore, fluorinating reagents are useful tools for the synthesis of pharmaceutical and agricultural compounds.

Fluorine-containing motifs are known to display unique chemical properties, especially in biological systems, and are crucial in the agricultural industry. Introduction of fluorine into bioactive compounds has been shown to be very effective in biological systems in regulating metabolism and also in drug discovery by improving the delivery and binding of pharmaceutically active agents to their specific targets. To facilitate the synthesis of fluorinated molecules, here in this book we have described a range of reagents and building blocks that are easier to handle than previous caustic sources of fluorine.

Chapter-2
Types of Fluorinating Reagents

2.1. Type of Fluorinating reagents:

There are two types of fluorinating reagents Electrophilic and Nucleophilic.

1) Electrophilic fluorination

Electrophilic fluorination is the combination of a carbon-centered nucleophile with an electrophilic source of fluorine to afford organofluorine compounds. Although elemental fluorine and reagents incorporating an oxygen-fluorine bond can be used for this purpose, they have largely been replaced by reagents containing a nitrogen-fluorine bond.[1]

Electrophilic fluorination offers an alternative to nucleophilic fluorination methods employing alkali or ammonium fluorides and methods employing sulfur fluorides for the preparation of organofluorine compounds. Development of electrophilic fluorination reagents has always focused on removing electron density from the atom attached to fluorine; however, compounds containing nitrogen-fluorine bonds have proven to be the most economical, stable, and safe electrophilic fluorinating agents. Electrophilic N-F reagents are either neutral or cationic and may possess either sp^2- or sp^3-hybridized nitrogen. Although the precise mechanism of electrophilic fluorination is currently unclear, highly efficient and stereoselective methods have been developed.

The most common fluorinating agents used for organic synthesis are given below.

❖ **Electrophilic fluorinating reagents**

- Inorganic fluorides: XeF_2, AgF_2, CoF_2 etc.
- NFSI
- Selectfluor

- N-Fluoropyridinium salt
- DFBA

2) Nucleophilic fluorination

The major alternative to electrophilic fluorination is, naturally, nucleophilic fluorination using reagents that are sources of "F⁻," for Nucleophilic displacement typically of chloride and bromide. Metathesis reactions employing alkali metal fluorides are the simplest.[9]

$$R_3CCl + MF \rightarrow R_3CF + MCl \ (M = Na, K, Cs)$$

Alkyl monofluorides can be obtained from alcohols and Olah reagent (pyridinium fluoride) or another fluoridating agents.

The decomposition of aryldiazonium tetrafluoroborates in the Sandmeyer[10] or Schiemann reactions exploit fluoroborates as F⁻ sources.

$$ArN_2BF_4 \rightarrow ArF + N_2 + BF_3$$

Although hydrogen fluoride may appear to be an unlikely nucleophile, it is the most common source of fluoride in the synthesis of organofluorine compounds. Such reactions are often catalysed by metal fluorides such as chromium trifluoride. 1,1,1,2-Tetrafluoroethane, a replacement for CFC's, is prepared industrially using this approach:[11]

$$Cl_2C=CClH + 4 \ HF \rightarrow F_3CCFH_2 + 3 \ HCl$$

Nucleophilic fluorination reactions of organic compounds using fluorinating reagents are one of the most widely used methodologies in the field of fluorine chemistry. DAST and SF_4 are widely utilized in one step reactions for the introduction of fluorine in organic compounds.

- ❖ **Nucleophilic fluorinating reagents**
  - Inorganic fluorides: AgF, CsF, KF, SbF_4, SF_4, HBF_4.
  - DAST
  - Morpho-DAST

- Deoxofluor

- Fluolead

- Perfluorobutanesulfonyl fluoride

- Phenofluor

- Xtalfluor-E & M

- Ishiwaka's reagent

- Yarovenko's reagent

- Triethylamine HF complex

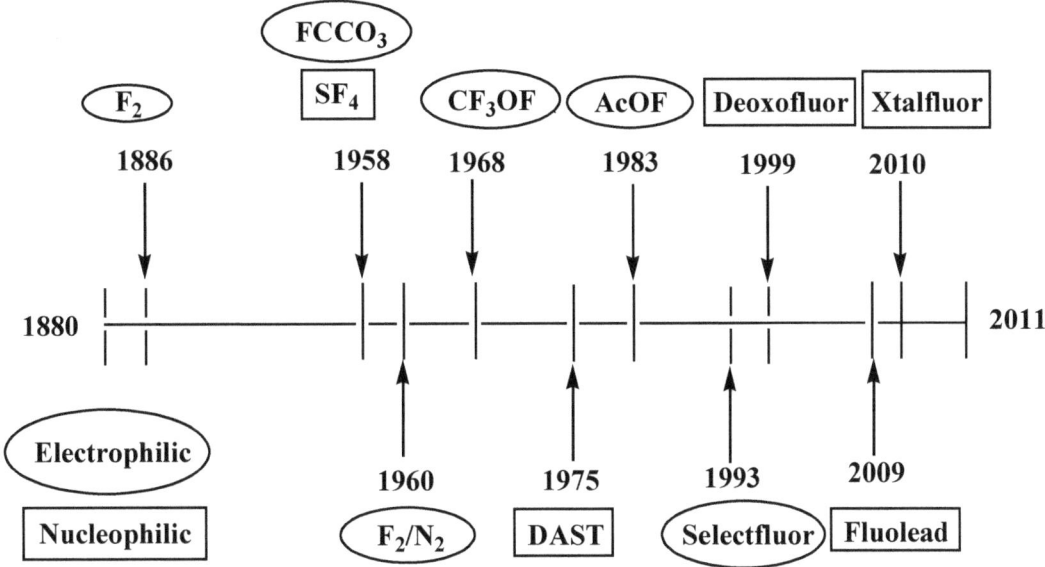

References:

1. Badoux, J.; Cahard, D. *Org. React.* **2007**, 69, 347. doi:10.1002/04712-64180.
2. Vogel, A. I.; Leicester, J.; Macey, W. A. T. "n-Hexyl Fluoride". *Org. Synth. Coll. Vol.* 4, p. 525.
3. Flood, D. T. "Fluorobenzene". *Org. Synth. Coll. Vol.* 2, p. 295.
4. William R. Dolbier, Jr. (**2005**). "Fluorine Chemistry at the Millennium". *Journal of Fluorine Chemistry.* 126 (2): 157. doi:10.1016/j.jfluchem. 2004.09. 033.

Chapter-3

Fluorinating Reagents

3.1. About Fluorinating reagents:

3.1.1. Accufluor

Accufluor® Fluorinating Agents are versatile, easy-to-use, and highly effective reagents for the introduction of fluorine into a wide variety of organic molecules. These stable solids require no special equipment or techniques in handling and will not attack glass like most fluoride containing reagents.

Structure:

IUPAC Name:	4-fluor-1-hydroxy-1,4-diazoniabicyclo[2.2.2]octane bis(tetrafluoroborate)
Other Name:	NFTh, Accufluor NFTh
CAS Number:	162241-33-0
Molecular formula:	$C_6H_{13}B_2F_9N_2O$
Molar mass:	321.79 g/mol
Appearance:	White solid
Density:	-
Melting point:	120 °C
Solubility in water:	-

Preparation and properties:

Accufluor is synthesized by the reaction of 1,4-diazabicyclo[2.2.2]octane-1-oxide with BF_3 and fluorine in acetonitrile.[1]

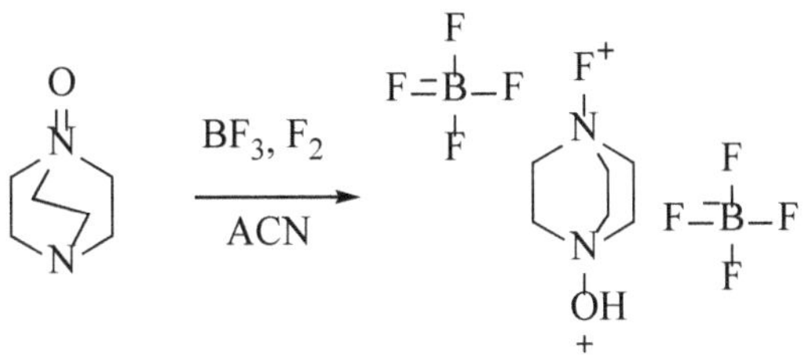

Scheme 1

Honeywell offers three different fluorinating agent products to cover a wide range of techniques for fluorine incorporation:

- NFTh, a strong fluorinating agent, selectively transfers fluorine to aromatic rings or electrone-rich olefins. In the presence of a Lewis acid, NFTh will convert the substrate to its mono- or difluoroderivatives.

- NFPy, a moderate strength agent, conveniently fluorinates organic sulfides and is particularly useful for enamides. NFSi is a mild, neutral reagent that is soluble in many common solvents such as tetrahydrofuran, methylene chloride, acetonitrile, and toluene.

- NFSi permits the incorporation of fluorine into a wide variety of nucleophilic substrates, including organometailic species and highly stabilized anions.

Application:

The use of 1-fluoro-4-hydroxy-1,4-diazoniabicyclo[2.2.2]octane bis(tetrafluoro-borate) (Accufluor NFTh) as a fluorine transfer reagent and methanol as solvent enabled direct regiospecific fluorofunctionalization of the α-carbonyl position in ketones without prior activation of the target molecules.[2][3]

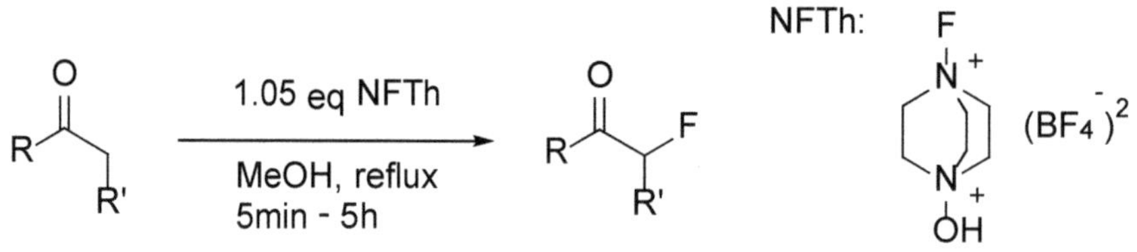

NFTh:

Scheme 2

Here we described fluorination yield using Accufluor.

	t	Yield %		t	Yield %
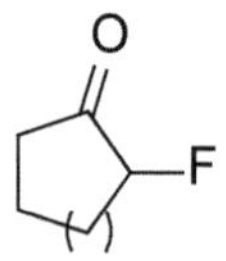		83	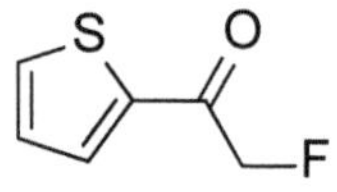	5 h	71
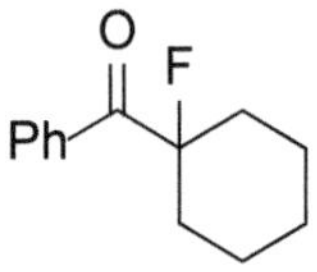		71	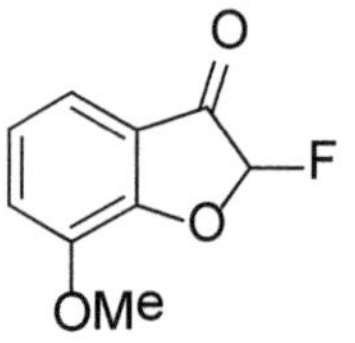	1˙5 h	81
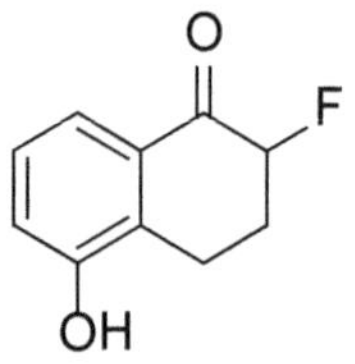	30 min	77	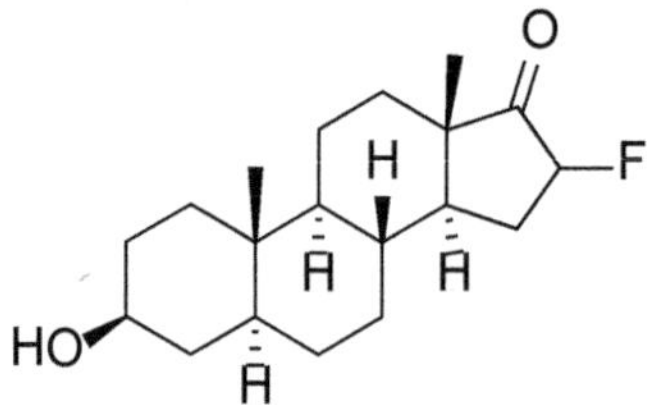	1˙5h	72

Safety:

Safety statement: S26-S36/37/39-S45-S61

Packing Group – III,   RIDADR: UN3335

Risk statement:- R22; R41; R43; R48/22; R50.

References:

1. *Tetrahedron Letters.* 1999, Vol. 40, # 14, p. 2673-2676.

2. S. Stavber, M. Jereb, M. Zupan. Direct α-Fluorination of Ketones Using N-F Reagents. *Synthesis,* 2002, 2609-2615. DOI: 10.1055/s-2002-35625.

3. G. Stavber, M. Zupan, S. Stavber, Micellar-System-Mediated Direct Fluorination of Ketones in Water. *Synlett*, 2009, 589-594.

3.1.2. Antimony trifluoride

Antimony trifluoride is the inorganic compound with the formula SbF_3. Sometimes called Swart's reagent, is one of two principal fluorides of antimony, the other being SbF_5. It appears as a white solid. As well as some industrial applications,[1] it is used as are agent in inorganic and organofluorine chemistry.

Structure:

IUPAC Name: Antimony (III) fluoride

Other Name: Trifluoroantimony

CAS Number: 7783-56-4

Molecular formula: SbF_3

Molar mass: 178.76 g/mol

Appearance: Light gray to white crystals

Odour: Pungent

Density: 4.379 g/cm^3

Melting point: 292 °C (558 °F; 565 K)

Boiling point: 376 °C (709 °F; 649 K)

Solubility in water: 385 g/100 mL (0 °C)
                     443 g/100 mL (20 °C)
                     562 g/100 mL (30 °C)

Solubility: Soluble in methanol, acetone
                               insoluble in ammonia

LD50 (Lethal dose):   7000 mg/kg

Preparation and properties:

In solid SbF_3, the Sb centres have octahedral molecular geometry and are linked by bridging fluoride ligands. Three Sb–F bonds are short (192 pm) and three are long (261 pm). Because it is a polymer, SbF_3 is far less volatile than related compounds AsF_3 and $SbCl_3$.[2]

SbF_3 is prepared by treating antimony trioxide with hydrogen fluoride:[3]

$$Sb_2O_3 + 6\ HF \rightarrow 2\ SbF_3 + 3\ H_2O$$

The compound is a mild Lewis acid, hydrolyzing slowly in water. With fluorine, it is oxidized to give antimony pentafluoride.

$$SbF_3 + F_2 \rightarrow SbF_5$$

Applications:

It is used as a fluorination reagent in organic chemistry.[4] This application was reported by the Belgium chemist Frederic Jean Edmond Swarts in 1892,[5] who demonstrated its usefulness for converting chloride compounds to fluorides. The method involved treatment with antimony trifluoride with chlorine or with antimony pentachloride to give the active species antimony trifluorodichloride ($SbCl_2F_3$). This compound can also be produced in bulk.[6] The Swarts reaction is generally applied to the synthesis of organofluorine compounds, but experiments have been performed using silanes.[7] It was once used for the industrial production off reon. Other fluorine-containing Lewis acids serve as fluorinating agents in conjunction with hydrogen fluoride.

SbF_3 is used in dyeing and in pottery, to make ceramic enamels and glazes.

Safety:

The lethal minimum dose (guinea pig, oral) is 100 mg/kg.[8]

References:

1. Sabina C. Grund, Kunibert Hanusch, Hans J. Breunig, Hans Uwe Wolf "Antimony and Antimony Compounds" in Ullmann's Encyclopedia of Industrial Chemistry, **2006**, Wiley-VCH, Weinheim. doi:10.1002/14356007 .a03_055.

2. Greenwood, Norman N.; Earnshaw, Alan (**1997**). Chemistry of the Elements (2nd ed.). Butterworth-Heinemann. ISBN 0080379419.

3. Handbook of Preparative Inorganic Chemistry, 2nd Ed. Edited by G. Brauer, Academic Press, **1963**, NY. Vol. 1. p. 199.

4. Tariq Mahmood and Charles B. Lindahl Fluorine Compounds, Inorganic, Antimony in Kirk-Othmer Encyclopedia of Chemical Technology. doi:10. 1002/0471238961.0114200913010813.a01.

5. Swarts (**1892**). *Acad. Roy. Belg 3*. (24): 474.

6. Patent US 4438088.

7. Booth, Harold Simmons; Suttle, John Francis (**1946**). "The Preparation and Fluorination of Dimethyl and Trimethyl Chlorosilanes". *J. Ac. Chem. Soc.* 68 (12): 2658-2660. doi:10.1021/ja01216a072.

8. Sabina C. Grund, Kunibert Hanusch, Hans J. Breunig, Hans Uwe Wolf "Antimony and Antimony Compounds" in Ullmann's Encyclopedia of Industrial Chemistry **2006**, Wiley-VCH, Weinheim. doi: 10.1002/14356007. a03_055.pub2.

3.1.3. Arsenic trifluoride

Arsenic trifluoride is a chemical compound of arsenic and fluorine with the formula AsF_3. It is a colourless liquid which reacts readily with water.[1]

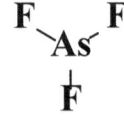

Structure:

IUPAC Name:	Arsenic (III) fluoride
Other Name:	Arsenic trifluoride, trifluoroarsane
CAS Number:	7784-35-2
Molecular formula:	AsF_3
Molar mass:	131.9168 g/mol
Appearance:	Colourless liquid
Density:	2.666 g/cm^3
Melting point:	-8.5 °C
Boiling point:	60.4 °C
Solubility in water:	Decomposes
Solubility:	Soluble in alcohol, ether, benzene and ammonia solution

Preparation and properties:

It can be prepared by reacting hydrogen fluoride, HF, with arsenic trioxide:[1]

$$6HF + As_2O_3 \rightarrow 2AsF_3 + 3H_2O$$

It has a pyramidal molecular structure in the gas phase which is also present in the solid.[1] In the gas phase the As-F bond length is 170.6 pm and the F-As-F bond angle 96.2°.[2]

Arsenic trifluoride is used as fluorinating non-metal chlorides to fluorides, in this respect it is less reactive than SbF_3.[1]

Salts containing AsF_4^- anion can be prepared for example $CsAsF_4$.[3] The potassium salt KAs_2F_7 prepared from KF and AsF_3 contains AsF_4^- and AsF_3 molecules with evidence of interaction between the AsF_3 molecule and the anion.[4]

With SbF_5 the ionic adduct AsF_2^+ SbF_6^- is produced [5]

Application:

Arsenic trifluoride application is described as below.[6]

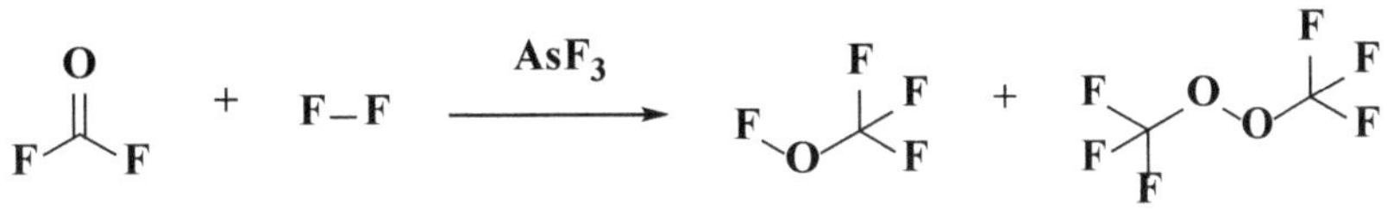

Safety:

A number of laboratory studies have reported mutagenic effects from this chemical. Use proper PPE to avoid the contact and breathing.

References:

1. Earnshaw, Alan (**1997**). *Chemistry of the Elements* (2nded.). Butterworth-Heinemann. ISBN 0080379419.

2. Wells A.F. (**1984**) *Structural Inorganic Chemistry* 5th edition Oxford Science Publications ISBN 0-19-855370-6.

3. New alkali metal and tetramethylammonium tetrafluoroarsenates(III), their vibrational spectra and crystal structure of cesium tetrafluoroarsenate (III)Klampfer P, Benkič P, Lesar A, Volavšek B, Ponikvar M , Jesih A., Collect. Czech. Chem. Commun. **2004**, 69, 339-350doi:10.1135/cccc200403-39.

4. Alkali-metal heptafluorodiarsenates(III): their preparation and the crystal structure of the potassium salt, Edwards A.J., Patel S.N., *J. Chem. Soc., Dalton Trans.* **1980**, 1630-1632, doi:10.1039/DT9800001630.

5. Fluoride crystal structures. Part XV. Arsenic trifluorideantimony pentafluoride, Edwards A. J., Sills R. J. C. *J. Chem. Soc. A.* **1971**, 942 - 945, doi:10.1039/ J19710000942.

6. Kennedy, R. C.; Cady, G. H. *Journal of Fluorine Chemistry.* **1973**, Vol. 3, p. 41-54.

3.1.4. Bromine pentafluoride

Bromine pentafluoride, BrF_5, is an inter halogen compound and a fluoride of bromine. It is a strong fluorination reagent.

It melts at −61.30 °C and boils at 40.25 °C. BrF_5 finds use in oxygen isotope analysis. Laser ablation of solid silicates in the presence of bromine pentafluoride releases O_2 for subsequent analysis.[1] It is also been tested as an oxidizer in liquid rocket propellants and is used as a fluorinating agent in the processing of uranium.

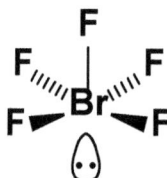

Structure:

IUPAC Name: Bromine pentafluoride

Other Name: Bromine fluoride

CAS Number: 7789-30-2

Molecular formula: BrF_5

Molar mass: 174.894 g/mol

Appearance: Pale yellow liquid

Density: 2.466 g/cm^3

Melting point: −61.30 °C (−78.34 °F; 211.85 K)

Boiling point: 40.25 °C (104.45 °F; 313.40 K)

Solubility in water: Reacts

Preparation and properties:

Bromine pentafluoride was first prepared in 1931 by the direct reaction of bromine with fluorine.[2] This reaction is suitable for the preparation of large quantities, and is carried out at temperatures over 150 °C (302 °F) with an excess of fluorine:

$$Br_2 + 5\,F_2 \rightarrow 2\,BrF_5$$

For the preparation of smaller amounts, potassium bromide is used: [2]

$$KBr + 3\,F_2 \rightarrow KF + BrF_5$$

This route yields bromine pentafluoride almost completely free of trifluorides and other impurities.[2]

Reactions:

Bromine pentafluoride reacts explosively with water, but when moderated by dilution with acetonitrile, it will form bromic acid and hydrofluoric acid, simple hydrolysis products:[3]

$$BrF_5 + 3\,H_2O \rightarrow HBrO_3 + 5\,HF$$

It is an extremely effective fluorinating agent, converting most uranium compounds to the hexafluoride at room temperature.

Application:

Bromine pentafluoride reacts with silyl compounds and metal chelate.[4]

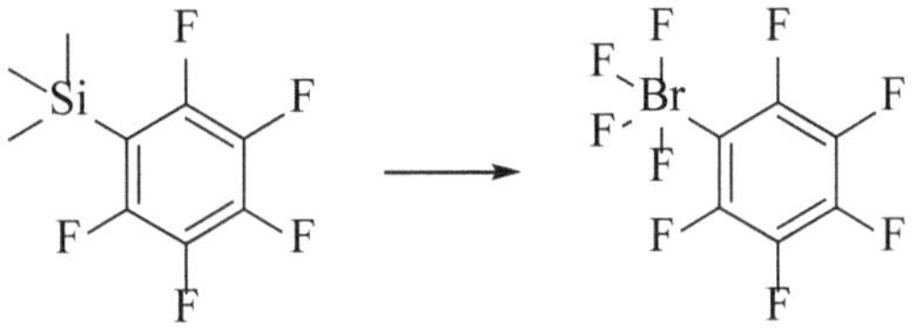

Scheme 1

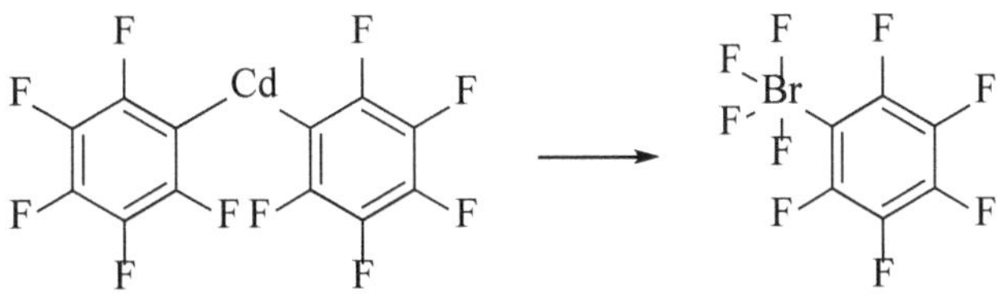

Scheme 2

Safety:

Bromine pentafluoride is severely corrosive to the skin, and its vapours are irritating to the eyes, skin, and mucous membranes. Exposure to 100 ppm for a few minutes is lethal to most experimental animals.

Chronic exposure may cause nephrosis and hepatosis.[4]

It may spontaneously ignite or explode upon contact with organic materials or metals in powdered form.[4]

References:

1. Clayton, R.; Mayeda, T. K. (**1963**). "The use of bromine pentafluoride in the extraction of oxygen from oxides and silicates for isotopic analysis". *Geochimica et Cosmochimica Acta.* 27 (1): 43–48. Bibcode: 1963GeCoA..27...43C. doi:10.1016/0016-7037(63)90071-1.

2. Hyde, G. A.; Boudakian, M. M. (**1968**). "Synthesis routes to chlorine and bromine pentafluorides". *Inorganic Chemistry.* 7 (12): 2648–2649. doi:10.1021/ic50070a039.

3. Greenwood, Norman N.; Earnshaw, Alan (**1997**). *Chemistry of the Elements* (2nd ed.). Butterworth-Heinemann. p. 834. ISBN 0080379419.

4. Brueuer Frohn, H. J. *Journal of Fluorine Chemistry.* **1990**, Vol. 47, # 2, p. 301-315.

5. Patnaik, Pradyot (**2007**). A comprehensive guide to the hazardous properties of chemical substances (3rd ed.). *Wiley Interscience. p. 480. ISBN 0-471-71458-5.*

3.1.5. Bromine trifluoride

Bromine trifluoride is an inter halogen compound with the formula BrF_3. It is a straw-coloured liquid with a pungent odour.[1] It is soluble in sulfuric acid but explodes on contact with water and organic compounds. It is a powerful fluorinating agent and an ionizing inorganic solvent. It is used to produce uranium hexafluoride (UF_6) in the processing and reprocessing of nuclear fuel.[2]

$$\begin{array}{c} F \\ \Big| \quad 1.72 \text{ A} \\ Br\text{---}F \\ 1.81 \text{ A} \quad \Big|\!\!\searrow_{\,86^{\,0}} \\ F \end{array}$$

Structure:	
IUPAC Name:	Bromine trifluoride
Other Name:	Bromine fluoride
CAS Number:	7787-71-5
Molecular formula:	BrF_3
Molar mass:	136.90 g/mol
Appearance:	Straw coloured liquid hygroscopic
Odour:	Choking, Pungent
Density:	2.803 g/cm^3
Melting point:	8.77 °C (47.79 °F; 281.92 K)
Boiling point:	125.72 °C (258.30 °F; 398.87 K)
Solubility in water:	Decomposes violently
Solubility:	Soluble in sulfuric acid

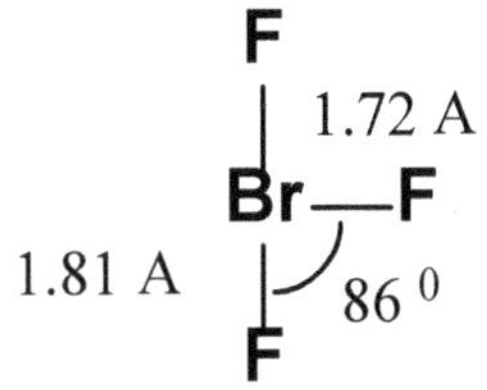

Preparation and properties:

Bromine trifluoride was first described by Paul Lebeau in 1906, who obtained the material by the reaction of bromine with fluorine at 20 °C:[3]

$$Br_2 + 3\ F_2 \rightarrow 2\ BrF_3$$

The disproportionation of bromine monofluoride also gives bromine trifluoride: [1]

$$3\ BrF \rightarrow BrF_3 + Br_2$$

Chemical properties:

BrF_3 is a fluorinating agent, but less reactive than ClF_3. The liquid is conducting, owing to auto ionisation:[2]

$$2\,BrF_3 \rightleftharpoons BrF_2^+ + BrF_4^-$$

Many ionic fluorides dissolve readily in BrF_3 forming fluoroanions:[2]

$$KF + BrF_3 \rightarrow KBrF_4$$

Application:

Reaction of Bromine trifluoride with trimethylsilyl compounds.[4]

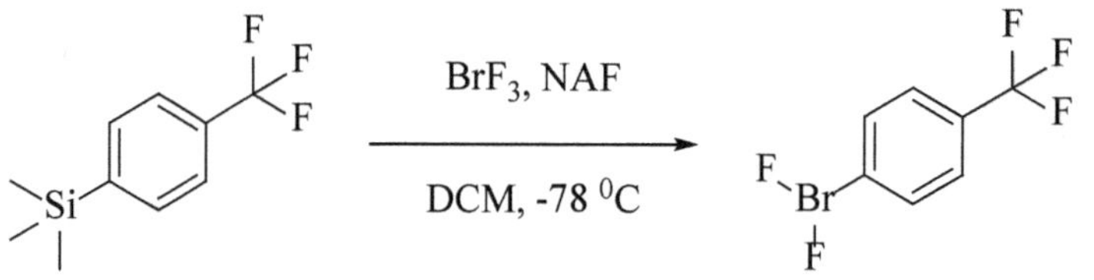

Scheme 1

Safety:

Contact with combustible material may cause fire. Contact with water liberates toxic, extremely flammable gas. This material increases the risk of fire and may aid combustion. Corrosive to the eyes, skin and respiratory system. Causes burns. Keep away from heat, sparks and flame. Keep away from water. Keep away from combustible material. Do not breathe gas. Do not breathe vapour or mist. Do not get in eyes or on skin or clothing. Use only with adequate ventilation. Keep container tightly closed and sealed until ready for use. Wash thoroughly after handling.

References:

1. Simons JH (**1950**). "Bromine (III) Fluoride Bromine Trifluoride". *Inorganic Synthesis*. 3: 184–186. doi:10.1002/9780470132340.ch48. ISBN 978-0-470-13234-0.

2. Greenwood, Norman N.; Earnshaw, Alan (**1997**). *Chemistry of the Elements* (2nd ed.). Butterworth-Heinemann. ISBN 0080379419.

3. Lebeau P. (**1906**). "The effect of fluorine on chloride and on bromine". *Annales de Chimie et de Physique* 9: 241–263.

4. *Journal of American Chemical Society.* **2003**, Vol. 125, # 50, p. 15304-15305.

3.1.6. Caesium Fluoride

Caesium fluoride or caesium fluoride is an inorganic compound usually encountered as a hygroscopic white solid. It is used in organic synthesis as a source of the fluoride anion. Caesium has the lowest electronegativity of all non-radioactive elements and fluorine has the highest electronegativity of all elements. It therefore has the strongest ionic compound bond discovered to be non-radioactive.

Structure:

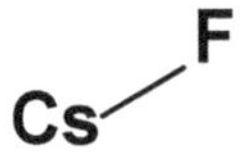

Cs — F

IUPAC Name: Caesium fluoride

Other Name: Cesium fluoride

CAS Number: 13400-13-0

Molecular formula: CsF

Molar mass: 151.90 g/mol

Appearance: white crystalline solid

Density: 4.115 g/cm^3

Melting point: 682 °C (1,260 °F; 955 K)

Boiling point: 1,251 °C (2,284 °F; 1,524 K)

Solubility in water: 322 g/100 mL (20 °C)

367 g/100 ml (18 °C)

Preparation and properties:

Caesium fluoride can be prepared by the reaction of caesium hydroxide (CsOH) with hydrofluoric acid (HF). The resulting salt can then be purified by recrystallization. The reaction is shown below:

$$CsOH(aq) + HF(aq) \rightarrow CsF(aq) + H_2O(l)$$

Another way to make caesium fluoride is to react caesium carbonate (Cs_2CO_3) with hydrofluoric acid. The resulting salt can then be purified by recrystallization. The reaction is shown below:

$$Cs_2CO_3(aq) + 2HF(aq) \rightarrow 2CsF(aq) + H_2O(l) + CO_2(g)$$

In addition, elemental fluorine and caesium can be used to form caesium fluoride as well, but doing so is very impractical because of the expense.[1] While this is not a normal route of preparation, caesium metal reacts vigorously with all the halogens to form caesium halides. Thus, it burns with fluorine gas, F_2, to form caesium fluoride, CsF according to the following reaction:

$$2Cs(s) + F_2(g) \rightarrow 2CsF(s)$$

CsF is more soluble than sodium fluoride or potassium fluoride. It is available in anhydrous form, and if water has been absorbed it is easy to dry by heating at 100 °C for two hours *in vacuo*.[2] CsF reaches a vapor pressure of 1 kilopascal at 825 °C, 10 kPa at 999 °C, and 100 kPa at 1249 °C.[3]

Application:

Being highly dissociated it is a more reactive source of fluoride than related salts. CsF is less hygroscopic alternative to tetra-n-butylammonium fluoride (TBAF) and TAS-fluoride (TASF) when anhydrous "naked" fluoride ion is needed.

As a base

As with other soluble fluorides, CsF is moderately basic, because HF is a weak acid. The low nucleophilicity of fluoride means it can be a useful base in organic chemistry.[4] CsF gives higher yields in Knoevenagel condensation reactions than KF or NaF.[6]

Formation of Cs-F bonds

Caesium fluoride is also a popular source of fluoride in organofluorine chemistry. For example, CsF reacts with hexafluoroacetone to form a caesium perfluoroalkoxide salt, which is stable up to 60 °C, unlike the corresponding sodium or potassium salt. It will convert electron-deficient aryl chlorides to aryl fluorides (halex reaction).[7]

Deprotection agent

Due to the strength of the Si–F bond, fluoride ion is useful for desilylation reactions (removal of Si groups) in organic chemistry; caesium fluoride is an excellent source of anhydrous fluoride for such reactions. Removal of silicon groups (desilylation) is a major application for CsF in the laboratory, as its anhydrous nature allows clean formation of water-sensitive intermediates. Solutions of caesium fluoride in THF or DMF attack a wide variety of organosilicon compounds to produce an organosilicon fluoride and a carbanion, which can then react with electrophiles,[5] for example:[6]

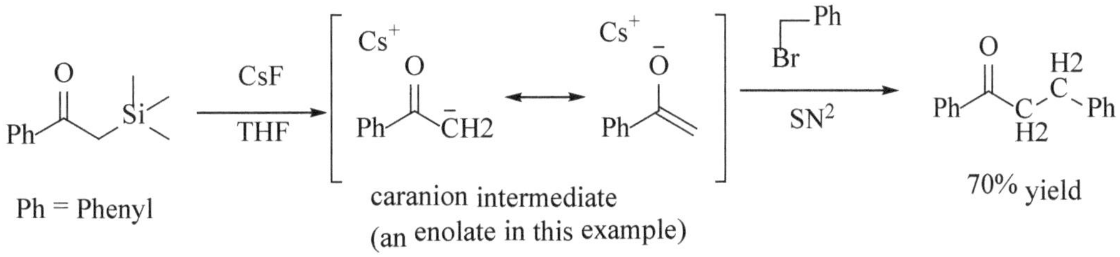

Ph = Phenyl

caranion intermediate
(an enolate in this example)

70% yield

Desilylation is also useful for the removal of silyl protecting groups.[8]

Other uses

Single crystals of the salt are transparent into the deep infrared. For this reason it is sometimes used as the windows of cells used for infrared spectroscopy.

Safety:

Like other soluble fluorides, CsF is moderately toxic.[9] Contact with acid should be avoided, as this forms highly toxic/corrosive hydrofluoric acid. The caesium ion (Cs^+) and caesium chloride are generally not considered toxic.[10]

References:

1. http://www.youtube.com/watch?v=TLOFaWdPxB0 Reacting Fluorine with Caesium.

2. Friestad, G. K.; Branchaud, B. P. (**1999**). Reich, H. J.; Rigby, J. H., ed. *Handbook of Reagents for Organic Synthesis: Acidic and Basic Reagents*. New York: Wiley. pp. 99–103.

3. http://www.physics.nyu.edu/kentlab/How_to/ChemicalInfo/VaporPressure/morepressure.pdf 6-63.

4. Greenwood, N.N.; Earnshaw, A. *Chemistry of the Elements*, Pergamon Press, Oxford, UK, **1984**.

5. Lide, D. R., ed. (**2005**). *CRC Handbook of Chemistry and Physics* (86th ed.). Boca Raton (FL): CRC Press. ISBN 0-8493-0486-5.

6. Fiorenza, M; Mordini, A; Papaleo, S; Pastorelli, S; Ricci, A (**1985**). "Fluoride ion induced reactions of organosilanes: the preparation of mono and dicarbonyl compounds from β-ketosilanes". *Tetrahedron Letters* 26 (6): 787. doi:10.1016/ S0040-4039(00)89137-6.

7. F. W. Evans, M. H. Litt, A. M. Weidler-Kubanek, F. P. Avonda (**1968**). "Reactions Catalyzed by Potassium Fluoride. 111. The Knoevenagel Reaction". *Journal of Organic Chemistry* 33 (5): 1837–1839. doi:10.1021/ jo01269a028.

8. Adam P. Smith, Jaydeep J. S. Lamba, and Cassandra L. Fraser (**2002**). "Efficient Synthesis of Halomethyl-2,2'-bipyridines: 4,4'-Bis(chloromethyl)-2,2'-bipyridine". *Org. Synth.* 78: 82.; *Coll. Vol.* 10, p. 107.

9. MSDS Listing for cesium fluoride. *www.hazard.com*. MSDS Date: April 27, 1993. Retrieved on September 7, **2007**.

10. "MSDS Listing for cesium chloride." *www.jtbaker.com*. MSDS Date: January 16, 2006. Retrieved on September 7, **2007**.

3.1.7. Chlorine monofluoride

Chlorine monofluoride is a volatile inter halogen compound with the chemical formula ClF. It is a colourless gas at room temperature and is stable even at high temperatures. When cooled to $-100\,°C$, ClF condenses as a pale yellow liquid. Many of its properties are intermediate between its parent halogens, Cl_2 and F_2.[1]

Structure:

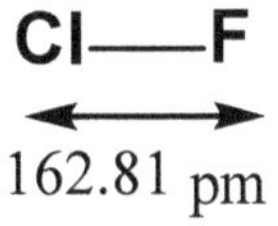

$$Cl\!-\!F$$
$$\longleftrightarrow$$
162.81 pm

IUPAC Name:	Chlorine monofluoride
Other Name:	Chlorine fluoride
CAS Number:	7790-89-8
Molecular formula:	ClF
Molar mass:	54.45 g/mol
Appearance:	Colourless gas
Density:	1.62 g/mL (liquid, $-100\,°C$)
Melting point:	$-155.6\,°C$ ($-248.1\,°F$; 117.5 K)
Boiling point:	$-100.1\,°C$ ($-148.2\,°F$; 173.1 K)
Solubility in water:	-

Preparation and properties:

Chlorine monofluoride synthesized by the reaction between chlorine and fluoride gas. [2] during reaction ClO_2 and ClF_3 generate as a by-products.

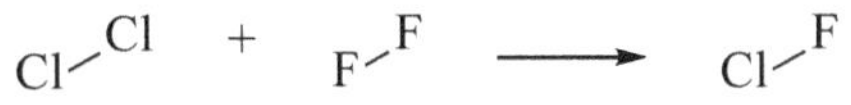

$$Cl\!-\!Cl \;+\; F\!-\!F \longrightarrow Cl\!-\!F$$

Chlorine trifluoride reacts with chlorine to give chlorine monofluoride. [3]

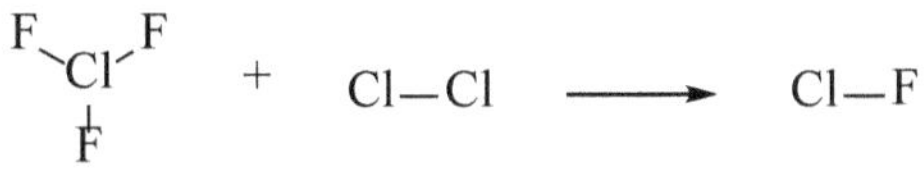

$$F\!-\!Cl\!-\!F \;(\,F\,) \;+\; Cl\!-\!Cl \longrightarrow Cl\!-\!F$$

Reaction of hypofluorous peroxyanhydride with chlorine gives chlorine monofluoride.[4]

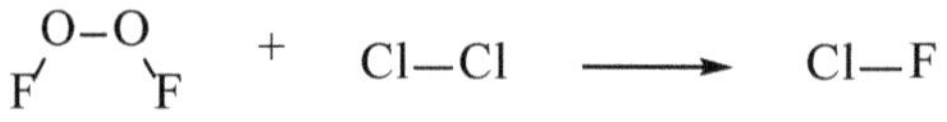

$$\underset{F}{\overset{O-O}{\diagup}}\underset{F}{\diagdown} \quad + \quad Cl-Cl \quad \longrightarrow \quad Cl-F$$

Reactivity:

Chlorine monofluoride is a versatile fluorinating agent, converting metals and non-metals to their fluorides and releasing Cl_2 in the process. For example, it converts tungsten to tungstenhexafluoride and selenium to seleniumtetrafluoride

$$W + 6\ ClF \rightarrow WF_6 + 3\ Cl_2$$

$$Se + 4\ ClF \rightarrow SeF_4 + 2\ Cl_2$$

ClF can also chlorofluorinate compounds, either by addition across a multiple bond or via oxidation. For example, it adds fluorine and chlorine across the triple bond of carbon monoxide:

$$CO + ClF \rightarrow \underset{F}{\overset{O}{\underset{\diagup}{\parallel}}}\underset{Cl}{\diagdown}$$

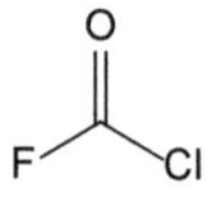

Safety:

To the best of our knowledge, the toxicological properties of this chemical have not been thoroughly investigated. Use appropriate procedures and precautions to prevent or minimize exposure.

References:

1. Otto Ruff, E. Ascher (**1928**). "Über ein neues Chlorfluorid-ClF$_3$". *Zeitschrift für anorganische und allgemeine Chemie.* 176 (1): 258-270. doi:10.1002/zaac.19281760121.

2. *Russian journal of Inorganic chemistry.* **1985**, Vol. 30, p. 1540-1541.

3. *Chemical physics Letters.* **1995**, Vol. 242, p. 407-414.

4. *Journal of American Chemical Society.* **1963**, vol. 85, p.1380-1385.

3.1.8. Chlorine pentafluoride

Chlorine pentafluoride is an interhalogen compound with formula ClF_5. This colourless gas is an strong oxidant that was once a candidate oxidizer for rockets. The molecule adopts a square pyramidal structure with C_{4v} symmetry,[1] as confirmed by its high resolution ^{19}F NMR spectrum.[2]

Structure:

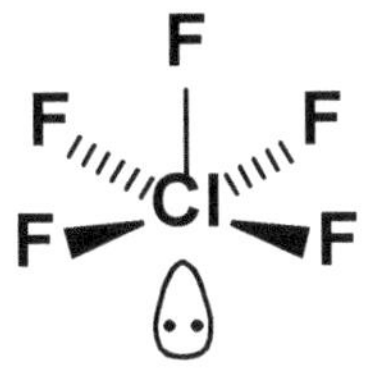

IUPAC Name:	Chlorine pentafluoride
Other Name:	Chlorine fluoride
CAS Number:	13637-63-3
Molecular formula:	ClF_5
Molar mass:	130.445 g/mol
Appearance:	Colourless gas
Density:	4.5 g/cm^3
Melting point:	−103 °C (−153 °F; 170 K)
Boiling point:	−13.1 °C (8.4 °F; 260.0 K)
Solubility in water:	Hydrolyze

Preparation and properties:

Some of the earliest research on the preparation was classified.[3][4] It was first prepared by fluorination of chlorine trifluoride at high temperatures and high pressures:

$$ClF_3 + F_2 \rightarrow ClF_5$$

NiF_2 catalyzes this reaction.[5]

Certain metal fluorides, $MClF_4$ (i.e. $KClF_4$, $RbClF_4$, $CsClF_4$) react with F_2 to produce ClF_5 and the corresponding alkali metal fluoride.[4]

Reactions:

In a highly exothermic reaction, water hydrolyses ClF_5 to produce chloryl -fluoride and hydrogen fluoride: [6]

$$ClF_5 + 2\ H_2O \rightarrow FClO_2 + 4\ HF$$

It is also a strong fluorinating agent. At room temperature it reacts readily with all elements except noble gases, nitrogen, oxygen and fluorine.[2]

Application:

Fluorination in 2,4,6-trifluoropyrimidine.[7]

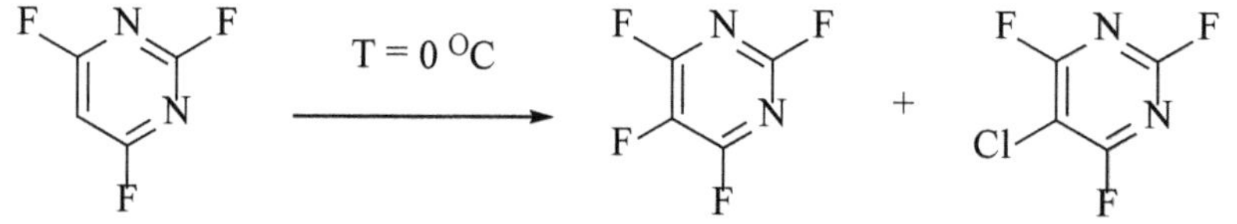

Scheme 1

Fluorination of tetrachloromethane gives three isomers.[7]

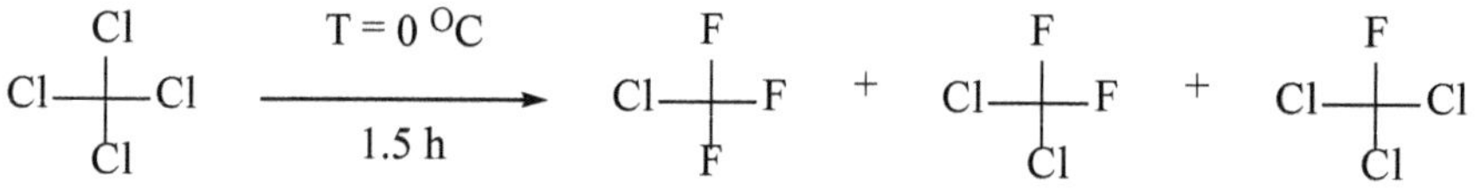

Scheme 2

Safety:

Wear suitable gloves, In case of insufficient ventilation, wear suitable respiratory equipment, Wear eye/face protection. In case of accident or if you feel unwell, seek medical advice immediately.

References:

1. Greenwood, Norman N.; Earnshaw, Alan (**1997**). *Chemistry of the Elements* (2nd ed.). Butterworth-Heinemann. p. 833. ISBN 0080379419.

2. Pilipovich, D., Maya, W., Lawton, E.A., Bauer, H.F., Sheehan, D. F., Ogimachi, N. N., Wilson, R. D., Gunderloy, F. C., Bedwell, V. E. (**1967**). "Chlorine pentafluoride. Preparation and Properties". *Inorganic Chemistry.* 6 (10): 1918. doi:10.1021/ic50056a036.

3. Clark, John (**1972**). *Ignition! An Informal History of Liquid Rocket Propellants.* Rutgers University Press. pp. 87–88. ISBN 0-8135-0725-1.

4. Smith D. F. (**1963**). "Chlorine Pentafluoride". *Science* 141 (3585): 1039–1040. doi:10.1126/science.141.3585.1039. PMID 17739492.

5. Šmalc, A., Žemva, B., Slivnik, J., and Lutar K. (**1981**). "On the Synthesis of Chlorine Pentafluoride". *Journal of Fluorine Chemistry.* 17 (4): 381–383. doi: 10.1016/S0022-1139(00)81783-2.

6. Greenwood, Norman N.; Earnshaw, Alan (**1997**). *Chemistry of the Elements* (2nd ed.). Butterworth-Heinemann. p. 834. ISBN 0080379419.

7. Boudakian, Max M. Hyde, Gene A. *Journal of Fluorine Chemistry.* **1984**, Vol. 25, p. 435-446.

3.1.9. Chlorine trifluoride

Chlorine trifluoride is an interhalogen compound with the formula ClF_3. This colourless, poisonous, corrosive, and extremely reactive gas condenses to a pale-greenish yellow liquid, the form in which it is most often sold (pressurized at room temperature). The compound is primarily of interest as a component in rocket fuels, in plasma less cleaning and etching operations in the semiconductor,[1][2] in nuclear reactor fuel processing,[3] and other industrial operations.[4]

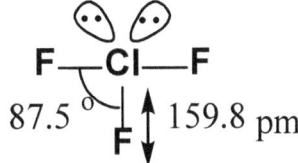

Structure:

IUPAC Name: Trifluoro-λ^3-chlorane

Other Name:	Chlorotrifluoride
CAS Number:	7790-91-2
Molecular formula:	ClF_3
Molar mass:	92.45 g/mol
Appearance:	Colourless gas or greenish-yellow liquid
Odour:	Sweet, Pungent, Irritating, Suffocating
Density:	4 mg/cm^3
Melting point:	−76.34 °C (−105.41 °F; 196.81 K)
Boiling point:	11.75 °C (53.15 °F; 284.90 K) (decomposes @ 180 °C)
Solubility in water:	Reacts violently

Preparation and properties:

It was first reported in 1930 by Ruff and Krug who prepared it by fluorination of chlorine; this also produced ClF and the mixture was separated by distillation.[5]

$$3 F_2 + Cl_2 \rightarrow 2 ClF_3$$

ClF_3 is approximately T-shaped, with one short bond (1.598 Å) and two long bonds (1.698 Å).[6] This structure agrees with the prediction of VSEPR theory, which predicts lone pairs of electrons as occupying two equatorial positions of a hypothetic trigonal bipyramid. The elongated Cl-F axial bonds are consistent with hypervalent bonding.

Pure ClF_3 is stable to 180 °C in quartz vessels; above this temperature it decomposes by a free radical mechanism to the elements.

Reactions:

Reaction with metal give chlorides and fluorides; phosphorus yields phosphorus trichloride (PCl_3) and phosphorus pentafluoride (PF_5); and sulfur yields sulfur dichloride (SCl_2) and sulfur tetrafluoride (SF_4). ClF_3 also reacts explosively with water, in which it oxidizes water to give oxygen or in controlled quantities, oxygen difluoride (OF_2), as well as hydrogen fluoride and hydrogen chloride. Metal oxides will react to form metal halides and oxygen or oxygen difluoride.

$$ClF_3 + 2H_2O \rightarrow 3HF + HCl + O_2$$

$$ClF_3 + H_2O \rightarrow HF + HCl + OF_2$$

The main use of ClF_3 is to produce uranium hexafluoride, UF_6, as part of nuclear fuel processing and reprocessing, by the fluorination of uranium metal:

$$U + 3\ ClF_3 \rightarrow UF_6 + 3\ ClF$$

Dissociates under the scheme:

$$ClF_3 \leftrightarrow ClF + F_2$$

Applications:

Military applications

Under the code name N-stoff ("substance N"), chlorine trifluoride was investigated for military applications by the Kaiser Wilhelm Institute in Nazi Germany from slightly before the start of World War II. Tests were made against mock-ups of the Line fortifications, and it was found to be an effective combined incendiary weapon and poison gas. From 1938, construction commenced on a partly bunkered, partly subterranean 31.76 km^2 munitions factory, the Falkenhagen industrial complex, which was intended to produce 50 tonnes of N-stoff per month, plus sarin. However, by the time it was captured by the advancing Red Army in 1945, the factory had produced only about 30 to 50 tonnes, at a cost of over 100 German Reichsmark per kilogram. N-stoff was never used in war.[7]

Semiconductor industry

In the semiconductor industry, chlorine trifluoride is used to clean chemical vapour deposition chambers.[8] It has the advantage that it can be used to remove semiconductor material from the chamber walls without having to dismantle the chamber.[8] Unlike most of the alternative chemicals used in this role, it does not need to be activated by the use of plasma since the heat of the chamber is enough to make it decompose and react with the semiconductor material.[8]

Rocket propellant

Chlorine trifluoride has been investigated as a high-performance storable oxidizer in rocket propellant systems. Handling concerns, however, prevented its use. John Drury Clark summarized the difficulties:

"It is, of course, extremely toxic, but that's the least of the problem. It is hypergolic with every known fuel, and so rapidly hypergolic that no ignition delay has ever been measured. It is also hypergolic with such things as cloth, wood, and test engineers, not to mention asbestos, sand, and water with which it reacts explosively. It can be kept in some of the ordinary structural metals steel, copper, aluminium, etc. because of the formation of a thin film of insoluble metal fluoride which protects the bulk of the metal, just as the invisible coat of oxide on aluminium keeps it from burning up in the atmosphere. If, however, this coat is melted or scrubbed off, and has no chance to reform, the operator is confronted with the problem of coping with a metal-fluorine fire. For dealing with this situation, I have always recommended a good pair of running shoes."[9][10][11]

Safety:

ClF_3 is a very strong oxidizing and fluorinating agent. It is extremely reactive with most inorganic and organic materials, including glass and teflon, and will initiate the combustion of many otherwise non-flammable materials without any ignition source. These reactions are often violent, and in some cases explosive. Vessels made from steel, copper or nickel resist the attack of the material due to formation of a thin layer of insoluble metal fluoride, but molybdenum, tungsten and titanium form volatile fluorides and are consequently unsuitable. Any equipment that comes into contact with chlorine trifluoride must be scrupulously cleaned and then passivated, because any contamination left may burn through the passivation layer faster than it can reform.

The ability to surpass the oxidizing ability of oxygen leads to extreme corrosivity against oxide-containing materials often thought as incombustible. Chlorine trifluoride and gases like it have been reported to ignite sand, asbestos, and other highly fire-retardant materials. In an industrial accident, a spill of 900 kg of chlorine trifluoride burned through 30 cm of concrete and 90 cm of gravel beneath.[12] Fire control/suppression is incapable of suppressing this oxidation, therefore the surrounding area is kept cool until the reaction ceases.[13] The compound reacts violently with water-based suppressors, and oxidizes in the absence of atmospheric oxygen, rendering atmosphere-displacement suppressors such as CO_2 and halon completely ineffective. It ignites glass on contact.[14]

Exposure of larger amounts of chlorine trifluoride, as a liquid or as a gas, ignites tissue. The hydrolysis reaction with water is violent and exposure results in a thermal burn. The products of hydrolysis are mainly hydrofluoric acid and hydrochloric acid, usually released as steam or vapour due to the highly exothermic nature of the reaction. Hydrofluoric acid is corrosive to human tissue, is absorbed through skin, selectively attacks bone, interferes with nerve function, and causes often-fatal fluorine poisoning. Hydrochloric acid is secondary in its danger to living organisms, but is several times more corrosive to most inorganic materials than hydrofluoric acid.

References:

1. Hitoshi Habuka, Takahiro Sukenobu, Hideyuki Koda, Takashi Takeuchi, and Masahiko Aihara (**2004**). "Silicon Etch Rate Using Chlorine Trifluoride". *Journal of the Electrochemical Society.* 151 (11): G783–G787. doi:10.1149/1.1806391.

2. United States Patent 5849092 "Process for chlorine trifluoride chamber cleaning".

3. Board on Environmental Studies and Toxicology, (BEST) (**2006**). *Acute Exposure Guideline Levels for Selected Airborne Chemicals: Volume 5 (citation at the National Academies Press).* Washington D.C.: National Academies Press. p. 40. ISBN 0-309-10358-4.

4. United States Patent 6034016 "Method for regenerating halogenated Lewis acid catalysts".

5. Otto Ruff, H. Krug (**1930**). "Über ein neues Chlorfluorid-CIF₃". *Zeitschrift für anorganische und allgemeine Chemie.* 190 (1): 270–276. doi:10.1002/ zaac. 19301900127.

6. Smith, D. F. (**1953**). "The Microwave Spectrum and Structure of Chlorine Trifluoride". *The Journal of Chemical Physics.* 21 (4): 609–614. Bibcode: 1953JChPh..21..609S.doi:10.1063/1.1698976.

7. "Bunker Tours" report on Falkenhagen.

8. "In Situ Cleaning of CVD Chambers". *Semiconductor International.* 6/1/1999.

9. Clark, John D. (**2001**). *Ignition!.* UMI Books on Demand.ISBN 0-8135-0725-1.

10. Clark, John D. (**1972**). *Ignition! An Informal History of Liquid Rocket Propellants.* Rutgers University Press. p. 214. ISBN 0-8135-0725-1.

11. ClF₃/Hydrazine at the Encyclopedia Astronautica.

12. Air Products Safetygram, http://web.archive.org/web/20060318221608, http:// www.airproducts.com/nr/rdonlyres/8479ed55-2170-4651-a3d4-223 b2957a9f3 /0/safetygram39.

13. "Chlorine Trifluoride Handling Manual". Canoga Park, CA: Rocketdyne. September **1961**. p. 24. Retrieved 2012-09-19.

14. Pradyot Patnaik (**2007**). *A comprehensive guide to the hazardous properties of chemical substances* (3rd ed.). Wiley-Interscience. p. 478.ISBN 0-471-71458-5.

3.1.10. Cobalt(III) fluoride

Cobalt(III) fluoride is the inorganic compound with the formula CoF_3. This highly reactive, hygroscopic brown solid is used to synthesize oregano - fluorine compounds.[1] CoF_3 is a powerful fluorinating agent that leaves CoF_2 as the by product.

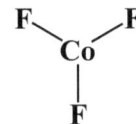

Structure:	
IUPAC Name:	Cobalt(III) trifluoride
Other Name:	Cobalt trifluoride
CAS Number:	10026-18-3
Molecular formula:	CoF_3
Molar mass:	115.928 g/mol
Appearance:	Brown powder
Density:	3.88 g/cm^3
Melting point:	927 °C (1,701 °F; 1,200 K)
Boiling point:	-
Solubility in water:	Reacts

Preparation and properties:

CoF_3 is prepared in the laboratory by treating $CoCl_2$ with fluorine at 250 °C:[2]

$$CoCl_2 + 3/2\ F_2 \rightarrow CoF_3 + Cl_2$$

This conversion is a redox reaction: Co^{2+} and Cl^- are oxidized to Co^{3+} and Cl_2, respectively, while F_2 is reduced to F^-. Cobalt (II) oxide (CoO) and cobalt (II) fluoride (CoF_2) can also be converted to cobalt (III) fluoride using fluorine.

Reactions:

CoF_3 decomposes upon contact with water to give oxygen:

$$4\ CoF_3 + 2\ H_2O \rightarrow 4\ HF + 4\ CoF_2 + O_2$$

CoF_3 is hygroscopic, forming a dihydrate (CAS#54496-71-8). It reacts with fluoride sources to give the anion $[CoF_6]^{3-}$, which is a rare example of a high-spin, octahedral cobalt (III) complex.

Applications:

Used as slurry, CoF_3 converts hydrocarbons to the per fluorocarbons:

$$2CoF_3 + R\text{-}H \rightarrow 2CoF_2 + R\text{-}F + HF$$

Such reactions are sometimes accompanied by rearrangements or other reactions. The related reagent $KCoF_4$ is more selective.[3]

Safety:

- DO NOT allow clothing wet with material to stay in contact with skin.

- " Avoid all personal contact, including inhalation.

- " Wear protective clothing when risk of exposure occurs.

References:

1. Coe, P. L. "Cobalt(III) Fluoride" in Encyclopedia of Reagents for Organic Synthesis (Ed: L. Paquette) **2004**, J. Wiley & Sons, New York.doi:10.1002/047084289X.rc185.

2. Priest, H. F. "Anhydrous Metal Fluorides" Inorganic Syntheses Mc Graw-Hill: New York, **1950**; Vol. 3, pages 171-183.doi:10.1002/9780470132340.ch47.

3. Coe, P. L. "Potassium Tetrafluorocobaltate(III)" in Encyclopedia of Reagents for Organic Synthesis (Ed: L. Paquette) **2004**, J. Wiley & Sons, New York. doi:10.1002/047084289X.rp251.

3.1.11. Cyanuric fluoride

Cyanuric fluoride or 2,4,6-trifluoro-1,3,5-triazine is a chemical compound with the formula $(CNF)_3$. It is a colourless, pungent liquid. It has been used as a precursor for fibre-reactive dyes, as a specific reagent for tyrosine residues in enzymes, and as a fluorinating agent.[1]

Structure:

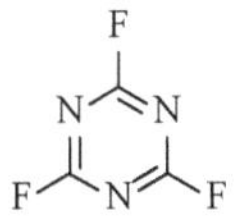

IUPAC Name: 2,4,6-trifluoro-1,3,5-triazine

Other Name: trifluorotriazine

CAS Number: 675-14-9

Molecular formula: $C_3F_3N_3$

Molar mass: 135.047 g/mol

Appearance: Colourless liquid

Density: 1.574 g/cm^3

Melting point: −38 °C (−36 °F; 235 K)

Boiling point: 74 °C (165 °F; 347 K)

Solubility in water: Reacts violently

Preparation and properties:

Cyanuric fluoride is prepared by fluorinating cyanuric chloride. The fluorinating agent may be SbF_3Cl_2,[2] KSO_2F,[3] or NaF.[4][5]

Cyanuric fluoride is used for the mild and direct conversion of carboxylic acids to acyl fluorides: [6]

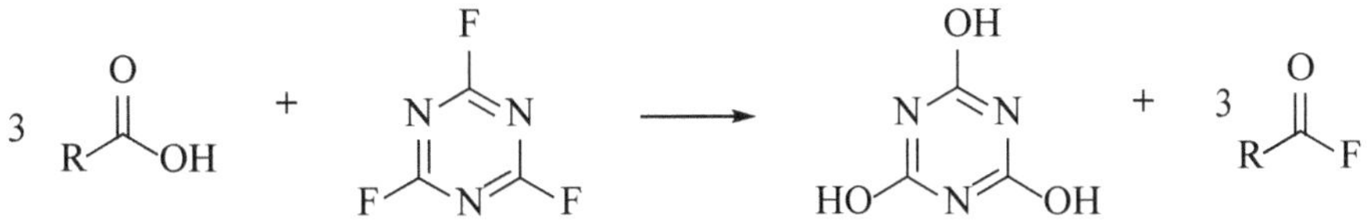

Scheme 1

Other fluorinating methods are less direct and may be incompatible with some functional groups.[7]

Cyanuric fluoride hydrolyses easily to cyanuric acid and it reacts more readily with nucleophiles than cyanuric chloride.[3] Pyrolysis of cyanuric fluoride at 1300 °C is a way to prepare cyanogen fluoride:[8]

$$(CNF)_3 \rightarrow 3 \ CNF.$$

Application:

The first method uses cyanuric fluoride as the fluorinating agent. In a typical experiment, equimolar amounts of the carboxylic acid, pyridine and cyanuric fluoride are mixed and stirred for 3-4 hours in dichloromethane at room temperature. At that time, ice-water is added to the reaction mixture and the precipitated cyanuric acid is filtered off. The organic phase is evaporated to dryness, which generally affords the pure acid fluoride in crystalline form.

Cyanuric fluoride convert carboxylic acid to carbonyl fluoride. Some examples are given below.[9][10][11] 3-phenyl-4-(trifluoromethyl)isoxazole-5-carboxylic acid converted to 3-phenyl-4-(trifluoromethyl)isoxazole-5-carbonyl fluoride.

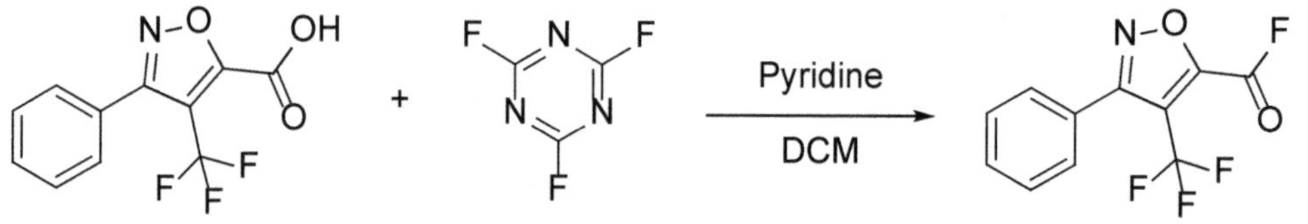

Scheme 2

1-(tetrahydro-2H-pyran-2-yl)-1H-pyrazole-4-carboxylic acid convert in to 1-(tetrahydro-2H-pyran-2-yl)-1H-pyrazole-4-carbonyl fluoride.[10]

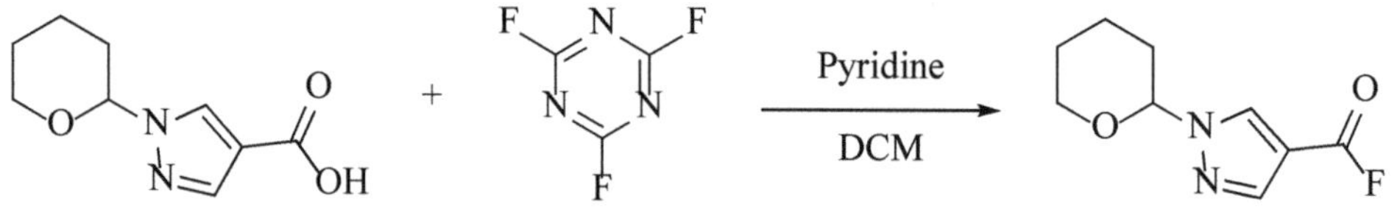

Scheme 3

Here in this example (R)-2-((N-(benzyloxy)formamido)methyl)hexanoic acid converted in to (R)-2-((N-(benzyloxy)formamido)methyl)hexanoyl fluoride.[11]

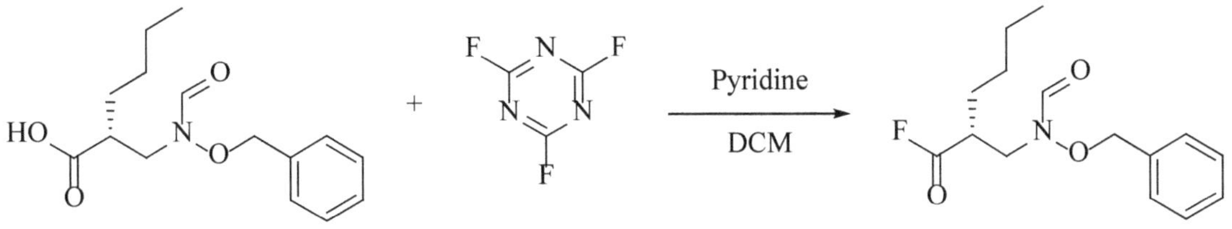

Scheme 4

(R)-4-(tert-butoxy)-2-(cyclopentylmethyl)-4-oxobutanoic acid converted to tert-butyl (R)-3-(cyclopentylmethyl)-4-fluoro-4-oxobutanoate.[11]

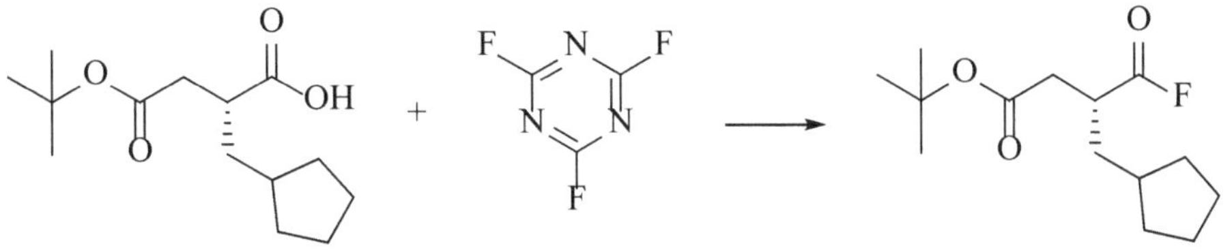

Scheme 5

Safety:

Inhalation (rat) LC50: 1276 ppm/1 h

Repeated or prolonged exposure to corrosives may result in the erosion of teeth, inflammatory and ulcerative changes in the mouth and necrosis (rarely) of the jaw. Bronchial irritation, with cough, and frequent attacks of bronchial pneumonia may ensue. Long-term exposure to respiratory irritants may result in disease of the airways involving difficult breathing and related systemic problems. Limited evidence suggests that repeated or long-term occupational exposure may produce cumulative health effects involving organs or biochemical systems. Extended exposure to inorganic fluorides causes fluorosis, which includes signs of joint pain and stiffness, tooth discoloration, nausea and vomiting, loss of appetite, diarrhea or constipation, weight loss, anemia, weakness and general unwellness. There may also be frequent urination and thirst.

Following acute or short term repeated exposure to hydrofluoric acid: "Subcutaneous injections of Calcium Gluconate may be necessary around the burnt area. Continued application of Calcium Gluconate Gel or subcutaneous Calcium

Gluconate should then continue for 3-4 days at a frequency of 4-6 times per day. If a "burning" sensation recurs, apply more frequently.

References:

1. "Fluorinated aromatic compounds". *Kirk-Othmer Encyclopedia of Chemical Technology* 11. Wiley-Interscience. **1994**, p. 608

2. Abe F. Maxwell, John S. Fry & Lucius A. Bigelow (**1958**). "The Indirect Fluorination of Cyanuric Chloride". *Journal of American Chemical Society* 80 (3): 548. doi:10.1021/ja01536a010.

3. Daniel W. Grisley, Jr, E. W. Gluesenkamp & S. Allen Heininger (**1958**). "Reactions of Nucleophilic Reagents with Cyanuric Fluoride and Cyanuric Chloride". *Journal of Organic Chemistry* 23 (11): 1802. doi:10.1021/jo01 105a620.

4. C. W. Tullock & D. D. Coffman (**1960**). "Synthesis of Fluorides by Metathesis with Sodium Fluoride". *Journal of Organic Chemistry* 25 (11) : 2016. doi:10. 1021/jo01081a050.

5. Steffen Groß, Stephan Laabs, Andreas Scherrmann, Alexander Sudau, Nong Zhang & Udo Nubbemeyer (**2000**). "Improved Syntheses of Cyanuric Fluoride and Carboxylic Acid Fluorides". *Journal für Praktische Chemie* 342 (7): 711. doi:10.1002/1521-3897(200009)342:7<711::AID-PRAC711>3.0.CO;2M.

6. George A. Olah, Masatomo Nojima & Istvan Kerekes (**1973**). "Synthetic Methods and Reactions; IV. Fluorination of Carboxylic Acids with Cyanuric Fluoride". *Synthesis.* **1973** (08): 487. doi:10.1055/s-1973-22238.

7. Barda, David A. (**2005**). "Cyanuric Fluoride". *Encyclopedia of Reagents for Organic Synthesis.* John Wiley & Sons. p. 77. doi:10.1002/ 047084289X. rn00043.

8. F. S. Fawcett & R. D. Lipscomb (**1964**). "Cyanogen Fluoride: Synthesis and Properties". *Journal of American Chemical Society.* 86 (13): 2576. doi:10.1021 /ja01067a011.

9. WO2011/59784A1, **2011**.

10. Millenium Pharmaceuticals Inc. WO2005/28474a2, **2005**.

11. Glaxosmithline. US2013/345120A1, **2013**.

3.1.12. Deoxofluor

Bis(2-methoxyethyl)aminosulfur Trifluoride (Deoxo-Fluor®): Originally reported by Lal and co-workers in 1999,[1] bis(2-methoxyethyl)aminosulfur trifluoride (Deoxo-Fluor®) has shown remarkable utility in organic synthesis as a thermally stable alternative to (diethylamino)sulfur trifluoride (DAST).

Structure:

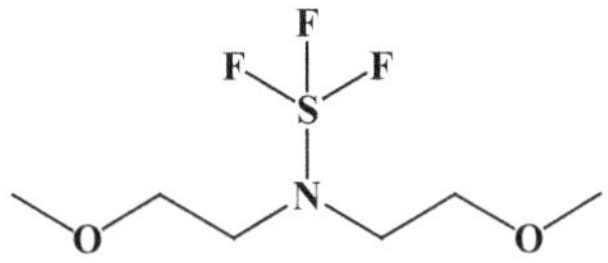

IUPAC Name: Bis(2-methoxyethyl)aminosulfur trifluoride

Other Name: DEOXO-FLUOR

CAS Number: 202289-38-1

Molecular formula: $C_6H_{14}F_3NO_2S$

Molar mass: 221.24 g/mol

Appearance: clear yellow liquid

Density: 1.2 g/mL at 25 °C(lit.)

Melting point: -

Boiling point: >80°C at 1.013 hPa - Decomposes on heating

Solubility in water: Reacts violently

Solubility: Organic non-protic solvents

Preparation and properties:

Deoxofluor is synthesized according to below scheme.

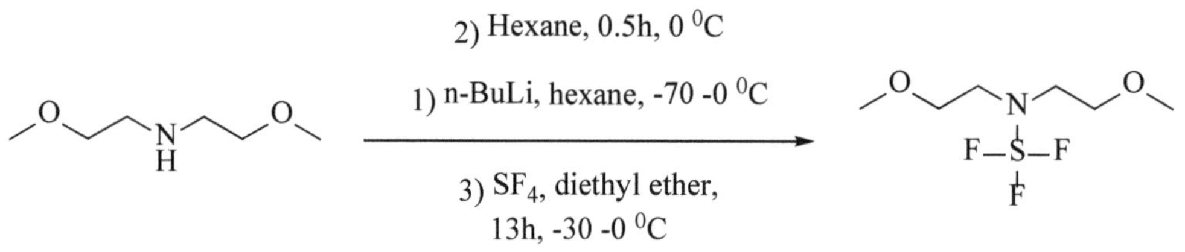

Scheme 1

According to reaction scheme, to the solution of bis(2-methoxyethyl)amine was added n-BuLi at -70 $^{\circ}$C. Then after react with SF_4 to yield Deoxofluor. [2]

Applications:

Deoxo-Fluor can easily convert alcohols to alkyl fluorides, aldehydes and ketones to gem-difluorides, and carboxylic acids to acid fluorides or trifluoromethyl derivatives (**Scheme 2**). [3]

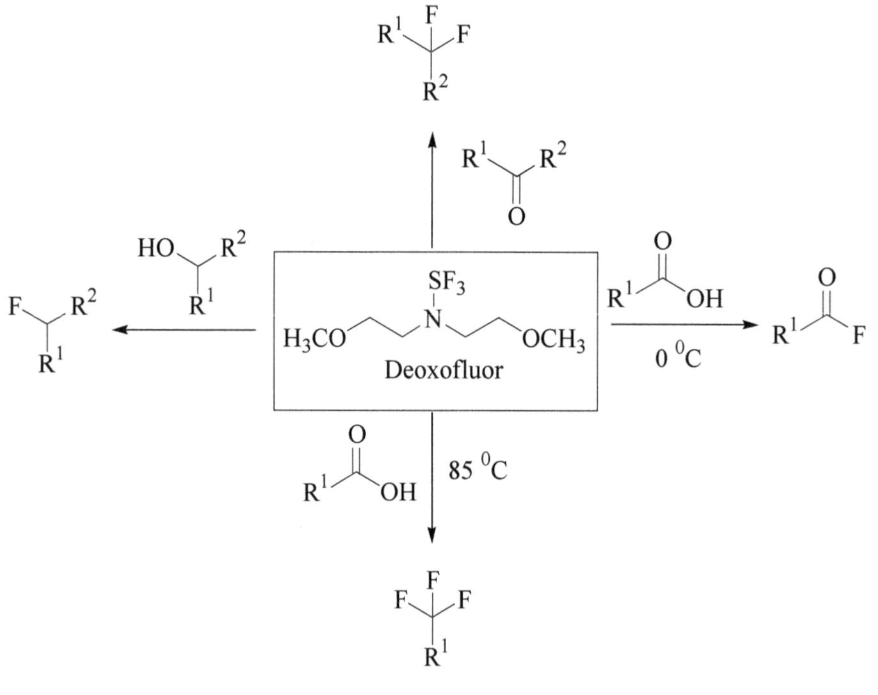

Scheme 2

A recent report showed the use of Deoxo-Fluor in the synthesis of a 2-azabicyclo[2.1.1]hexane analogue of 4-fluoroproline (**Scheme 3**).[4] Although the reaction generally proceeds with inversion at the chiral carbon, the configuration is retained in this constrained substrate due to neighbouring group participation of the amide group.

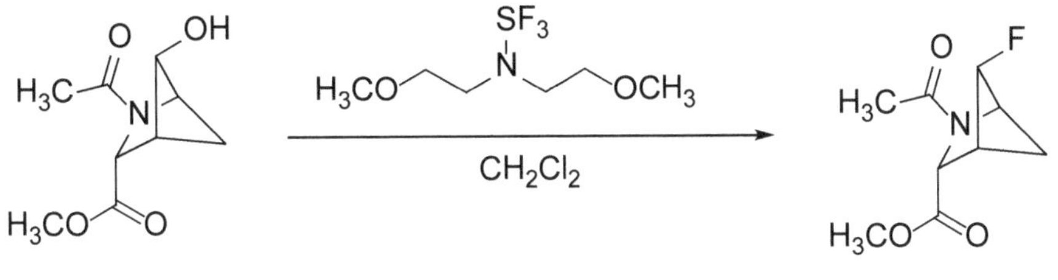

Scheme 3

Deoxo-Fluor can be used to effect a wide array of additional transformations. One recent example highlights its use in the syntheses of a series of chiral C_2 bis-oxazoline ligands (**Scheme 4**).[5] The chemoselectivity of Deoxo-Fluor was superior to other reagents, including DAST, and was tolerant of steric and electronic variations within the ligands. Furthermore, the Deoxo-Fluor protocol also allowed the majority of the ligands to be purified without requiring chromatography.

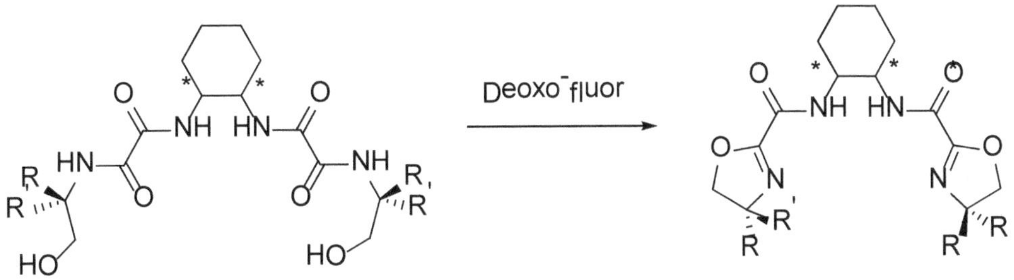

Scheme 4

Gunda Georg at the University of Kansas has reported the use of Deoxo-Fluor in a one-flask protocol to convert acids to amides and peptides or Weinreb amides (**Scheme 5**).[6] The reaction proceeded under mild conditions and provided the desired products in high yields with facile purification.

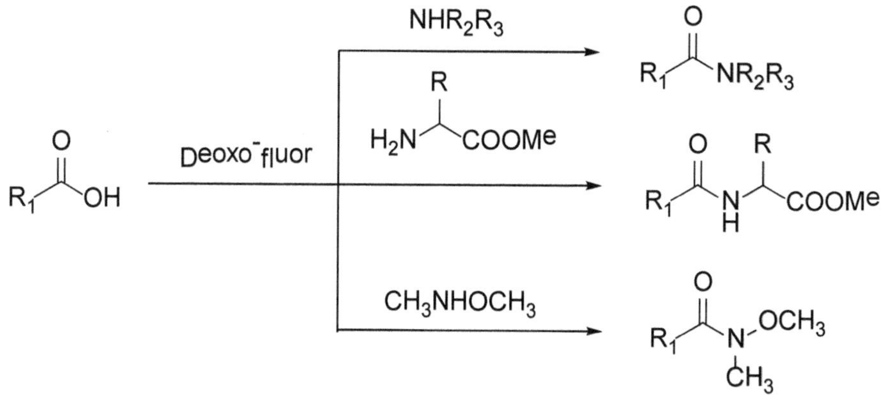

Scheme 5

The Kangani group has devoted several publications to one-pot transformations achievable with Deoxo-Fluor.[7] Generally, the reactions begin with the conversion of a carboxylic acid to an acid fluoride, which is then reacted with various nucleophiles to produce oxazolines, aldehydes, ketones, benzoxazoles, oxadiazoles, acyl azides or nitriles (**Scheme 6**).

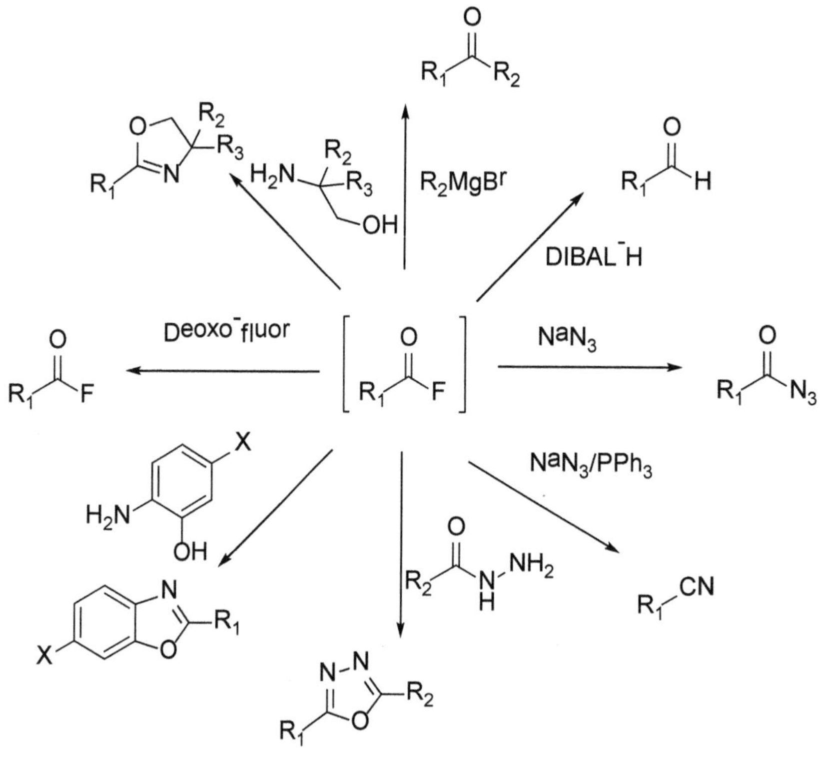

Scheme 6

Safety:

Material reacts with moisture on the skin, eyes, and mucous membranes to generate hydrogen fluoride. Hydrogen fluoride is extremely destructive and may cause deep progressive burns that induce subcutaneous tissues to become blanched and bloodless resulting in lesions of dead tissue that are slow to heal.

Consult a physician. Show this safety data sheet to the doctor in attendance. Hydrofluoric (HF) acid burns require immediate and specialized first aid and medical treatment. Symptoms may be delayed up to 24 hours depending on the concentration of HF. After decontamination with water, further damage can occur due to penetration/absorption of the fluoride ion. Treatment should be directed toward binding the fluoride ion as well as the effects of exposure. Skin exposures can be treated with a 2.5% calcium gluconate gel repeated until burning ceases. More serious skin exposures may require subcutaneous calcium gluconate except for digital areas unless the physician is experienced in this technique, due to the potential for tissue injury from increased pressure. Absorption can readily occur through the subungual areas and should be considered when undergoing decontamination. Prevention of absorption of the fluoride ion in cases of ingestion can be obtained by giving milk, chewable calcium carbonate tablets or Milk of Magnesia to conscious victims. Conditions such as hypocalcaemia, hypomagnesaemia and cardiac arrhythmias should be monitored for, since they can occur after exposure.

References:

1. (a) Lal, G. S. et al. *J. Org. Chem.* **1999**, *64,* 7048. (b) Lal, G. S. et *al. Chem. Commun.* **1999**, 215.
2. *J. Org. Chem.* **1999**, vol.64, #19, p. 7048-7054.
3. Singh, R. P.; Shreeve, J. M. *Synthesis* **2002**, 2561.
4. Jenkins, C. L. et al. *J. Org. Chem.* **2004**, *69,* 8565.
5. Albano, V. G. et al. *J. Org. Chem.* **2006**, *71,* 6451.
6. White, J. M. et al. *J. Org. Chem.* **2004**, *69,* 2573.

7. (a) Kangani, C. O.; Kelley, D. E. *Tetrahedron Lett.* **2005**, *46*, 8917. (b) Kangani, C. O. et al. *Tetrahedron Lett.* **2006**, *47*, 6289. (c) Kangani, C. O. et al. *Tetrahedron Lett.* **2006**, *47*, 6497. (d) Kangani, C. O. et al. *Tetrahedron Lett.* **2007**, *48*, 5933.

3.1.13. Diethylaminosulfur trifluoride (DAST)

Diethylaminosulfur trifluoride (DAST) is the organosulfur compound with the formula Et_2NSF_3. This liquid is a fluorinating reagent used for the synthesis of organofluorine compounds. The compound is colourless; older samples assume an orange colour.

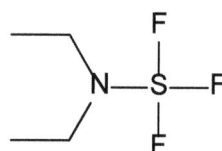

Structure:

IUPAC Name:	*N,N*-Diethylaminosuflur trifluoride
Other Name:	diethyl(trifluorosulfido)amine
CAS Number:	**38078-09-0**
Molecular formula:	$C_4H_{10}F_3NS$
Molar mass:	161.19 g/mol
Appearance:	Colourless oil
Density:	1.220 g/cm^3
Melting point:	-
Boiling point:	30 to 32 °C (86 to 90 °F; 303 to 305 K) at 3 mmHg
Solubility in water:	Reacts with water
Solubility:	soluble in acetonitrile

Preparation and properties:

DAST is prepared by the reaction of diethylaminotrimethylsilane and sulfur tetrafluoride:[1]

$$Et_2NSiMe_3 + SF_4 \rightarrow Et_2NSF_3 + Me_3SiF$$

The Organic Syntheses protocol calls for trichlorofluoromethane as a solvent, a compound that has been banned under the Montreal Protocol and is no longer available as a commodity chemical. Diethyl ether may be used instead with no decrease in yield.[2]Because of the dangers involved in the preparation of DAST (glass etching, possibility of exothermic events), it is often purchased from a commercial source. At one time Carbolabs [3] was one of the few suppliers of the chemical but a number of companies now sell DAST.

Applications:

Use in organic synthesis: DAST converts alcohols to the corresponding alkyl fluorides as well as aldehydes and unhindered ketones to geminal difluorides. Carboxylic acids react no further than the acyl fluoride (sulfur tetrafluoride effects the transformation $-CO_2H \rightarrow -CF_3$). DAST is used in preference to the more classical gaseous SF4, since as a liquid it is more easily handled. Acid-labile substrates are less likely to undergo rearrangement and elimination since DAST is less prone to contamination with acids. Reaction temperatures are milder as well - alcohols typically react at -78 °C and ketones around 0 °C.

Fluorinations with DAST can be carried out in conventional glass equipment, although etching of the glass may result from reaction byproducts. Reactions are typically carried out in aprotic or non-polar solvents. Moisture and atmospheric oxygen should be excluded from the reaction. Reactions are generally started at -78 °C and warmed to room temperature or above; however, reactions should not be heated above 80°C, as decomposition of the fluorinating reagent begins to occur at this temperature. Workup usually involves pouring the reaction mixture over water or ice, followed by neutralization of acidic byproducts with sodium bicarbonate. Standard purification methods can be used to isolate the desired fluorinated products.

Aldehydes and ketones react with DAST to form the corresponding geminal difluorides. Fluorination of enolizable ketones gives a mixture of the difluoroalkane and vinyl fluoride. In <u>glyme</u> with fuming sulfuric acid, the vinyl fluoride product predominates.[4] Electron-rich carbonyl compounds, such as esters and amides, do not react with DAST or other aminosulfuranes.

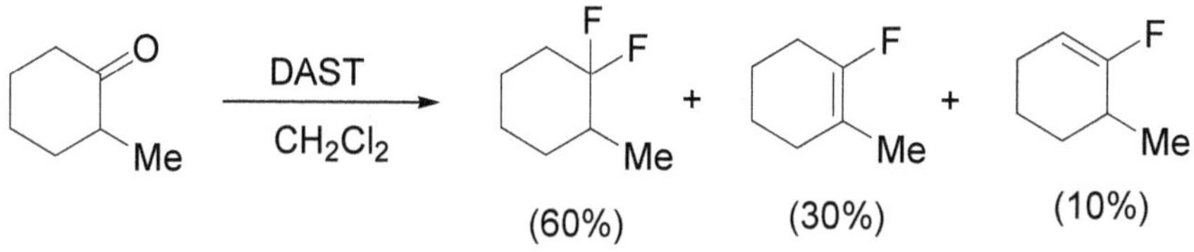

Scheme 1

Epoxides may yield a variety of products depending on their structure. Generally, the products that form in highest yield are vicinal difluorides and bis(α-fluoroalkyl)ethers. However, this reaction results in low yields and is not synthetically useful.[5]

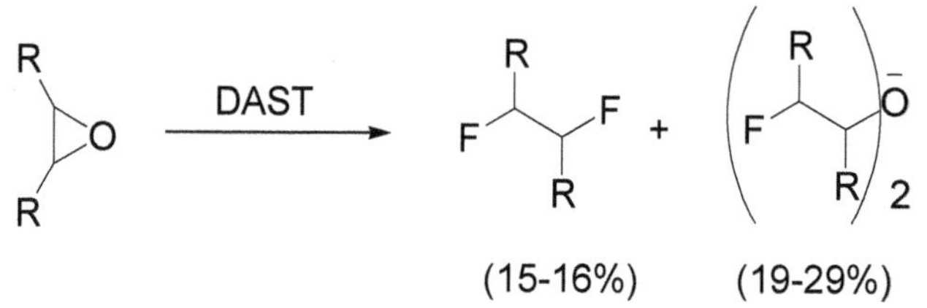

Scheme 2

The polar mechanism of fluorination by DAST implies that certain substrates may suffer Wagner-Meerwein rearrangements. This process has been observed in the fluorination of pivalaldehyde, which affords a mixture of 1,2-difluoro-1,2-dimethylpropane, 1,1-difluoro-2,2-dimethylpropane, and 1-fluoro-2,2-dimethylethylene.[6]

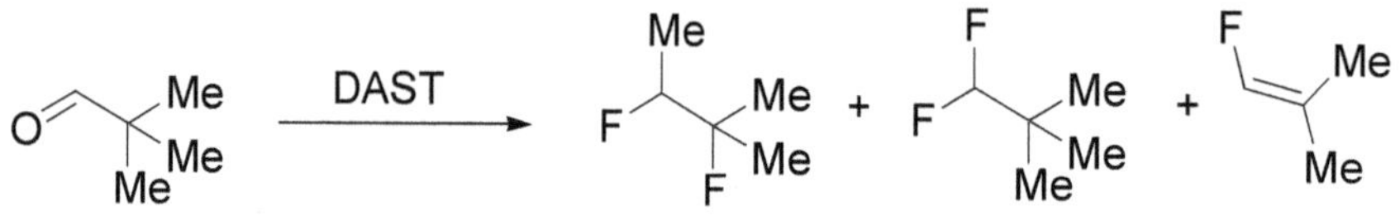

Scheme 3

Diols can undergo pinacol rearrangement under fluorination conditions.[7]

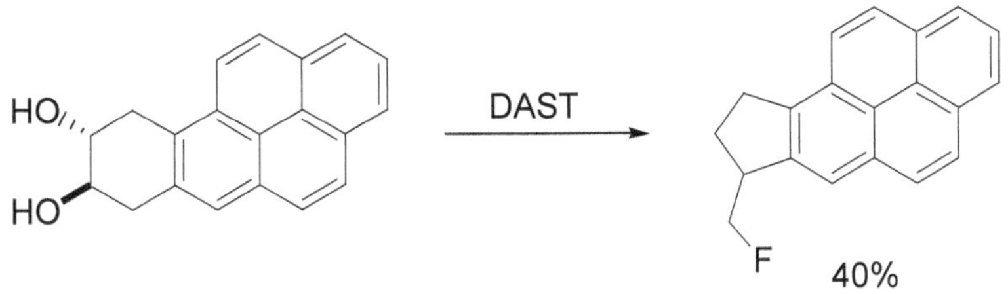

Scheme 4

When sulfoxides are treated with DAST, an interesting Pummerer-type rearrangement occurs to afford α-fluoro sulfides.[8]

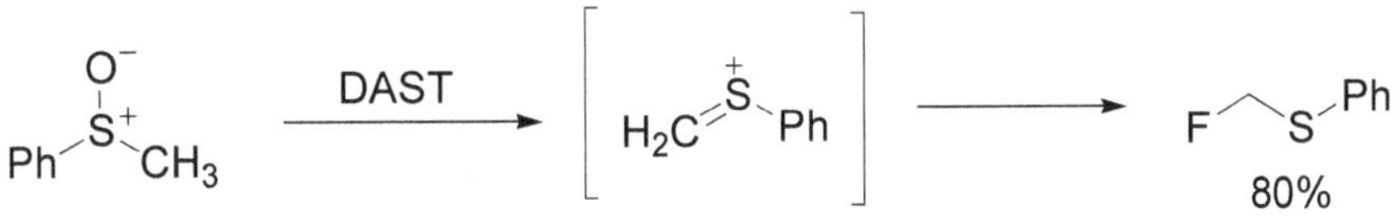

Scheme 5

Safety:

Upon heating, DAST converts to SF_4 and $(NEt_2)_2SF_2$, a high-boiling and explosive compound. To minimize accidents, samples are maintained below 50 °C. Bis-(2-methoxyethyl)aminosulfur trifluoride is more thermally robust.

Recent alternatives [9][10] have been manufactured by Omega Chem based on DAST but crystalline and with better handling properties. The XtalFluor range [11] offers XtalFluor-E (diethylamine) and XtalFluor-M (morpholine) difluoro-sulfonium salts (tetrafluoroborates).

Corrosive, flammable, can be explosive.

References:

1. W. J. Middleton, E. M. Bingham "Diethylaminosulfur Trifluoride" *Organic Syntheses,* Coll. Vol. 6, p.440; Vol. 57, p.50. Online version.

2. L. N. Markovskij, V. E. Pashinnik, and A. V. Kirsanov (**1973**). *Synthesis* (787).

3. Reaction of sulfoxides with diethylaminosulfur trifluoride: Fluoromethyl phenyl sulfone, Areagent for the synthesis of fluoroalkenes, *Organic Syntheses*, Coll. Vol. 9, p.446 (**1998**); Vol. 72, p.209 (1995).

4. Boswell, Jr., G. A. U.S. Patent 4212815 (**1980**) [C.A., 93, 239789w (1980)].

5. Hudlický, M. *J. Fluorine Chem.* **1987**, *36*, 373.

6. Middleton, J. *J. Org. Chem.* **1975**, *40*, 574.

7. Newman, S.; Khanna, M.; Kanakarajan, K. *J. Am. Chem. Soc.* **1979**, *101*, 6788.

8. McCarthy, R.; Peet, P.; LeTourneau, E.; Inbasekaran, M. *J. Am. Chem. Soc.* **1985**, 107, 735.

9. l'Heureux, A.; Beaulieu, F.; Bennett, C.; Bill, D. R.; Clayton, S.; Laflamme, F. O.; Mirmehrabi, M.; Tadayon, S.; Tovell, D.; Couturier, M. (**2010**). "Aminodifluorosulfinium Salts: Selective Fluorination Reagents with Enhanced Thermal Stability and Ease of Handling". *The Journal of Organic Chemistry.* **75** (10): 3401. doi:10.1021/jo100504x.

10. Beaulieu, F.; Beauregard, L. P.; Courchesne, G.; Couturier, M.; Laflamme, F. O.; l'Heureux, A. (**2009**). "Aminodifluorosulfinium Tetrafluoroborate Salts as Stable and Crystalline Deoxofluorinating Reagents". *Organic Letters* 11 (21): 5050. doi:10.1021/ol902039q.

11. http://www.omegachem.com/index.php?option=com_content&view=article&id=53.

3.1.14. 2,2-difluoro-1,3-dimethylimidazolidine

2,2-Difluoro-1,3-dimethylimidazolidine (DFI) is a new deoxo-fluorinating agent that is useful for the conversion of alcohols to monofluorides, and aldehydes/ketones to *gem*-difluorides under mild conditions.

Structure:

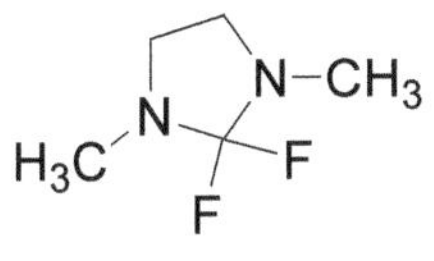

IUPAC Name: 2,2-difluoro-1,3-dimethylimidazolidine

Other Name: DFI

CAS Number: 220405-40-3

Molecular formula: $C_5H_{10}F_2N_2$

Molar mass: 136.14 g/mol

Appearance: Liquid

Density: 1.096 g/mL at 25 °C

Melting point: -8.7 °C

Boiling point: 47 °C at 4.9 kPa

Solubility in water: -

Solubility: Soluble in n-Hexane, ether, acetonitrile, toluene, chloroform, dichloromethane, and dioxane.

Refractive Index: n^{20}_D 1.40

Preparation and properties:

Reaction of 2,2-dichloro-1,3-dimethylimidazolidine with potassiumfluoride in acetonitrile gives 2,2-difluoro-1,3-dimethylimidazolidine.[1]

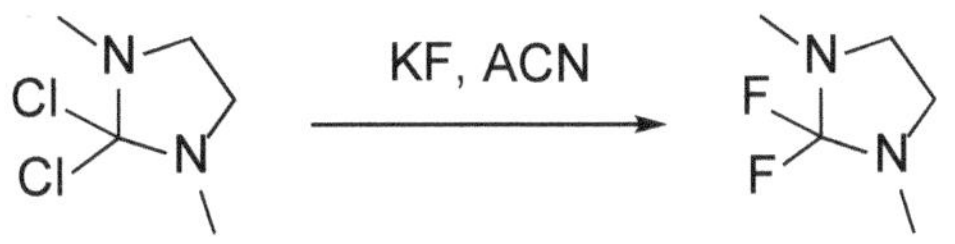

Scheme 1

2,2-difluoro-1,3-dimethylimidazolidine also synthesized from 2-chloro-1,3-dimethylimidazolidine.[2]

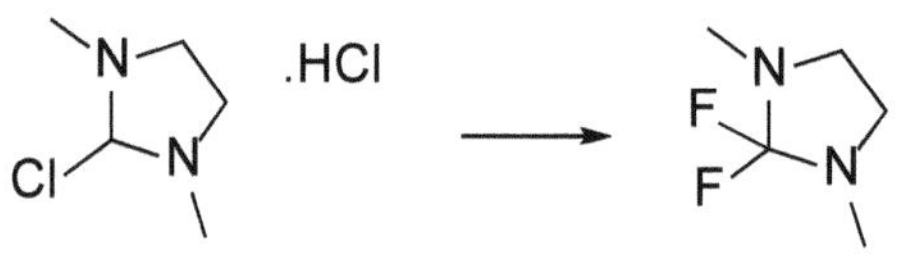

Scheme 2

Using Helaxe like reaction dichloro compound converted in to difluoro compound.[3]

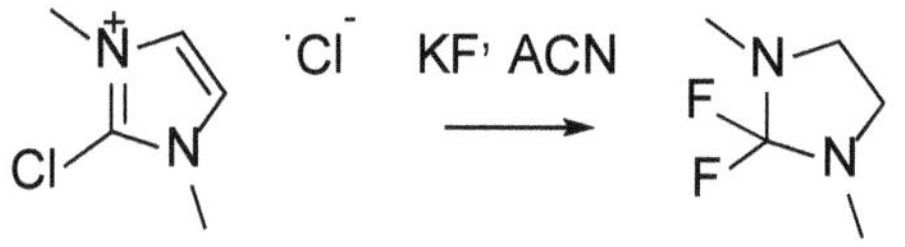

Scheme 3

Applications:

Some organic fluorinating compound are synthesized by the using 2,2-difluoro-1,3-dimethylimidazolidine.[3][4]

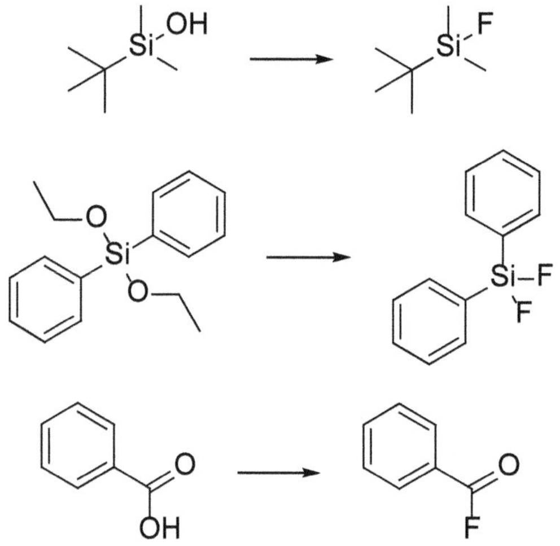

Scheme 4

Safety:

If breathed in, move person into fresh air. If not breathing, give artificial respiration. Consult a physician. Take off contaminated clothing and shoes immediately. Wash off with soap and plenty of water. Consult a physician. Do NOT induce vomiting. Never give anything by mouth to an unconscious person. Rinse mouth with water.

Rinse thoroughly with plenty of water for at least 15 minutes and consult a physician.

Use personal protective equipment. Avoid breathing vapours, mist or gas. Ensure adequate ventilation. Remove all sources of ignition. Evacuate personnel to safe areas. Beware of vapours accumulating to form explosive concentra-tions. Vapours can accumulate in low areas.

References:

1. WO 2005/75413 A1, **2005**.

2. EP1813596 A1, **2007**.

3. US6344579 B1, **2002**.

4. US6329529 B1, **2001**.

3.1.15. Difluoromethyl phenyl sulfone

A facile and efficient nucleophilic difluoromethylation of primary alkyl halides has been disclosed through a novel nucleophilic substitution–reductive desulfonylation strategy, using difluoromethyl phenyl sulfone as a difluoromethyl anion ("CF_2H") equivalent.

Structure:	
IUPAC Name:	Difluoromethyl Phenyl Sulfone
Other Name:	[(Difluoromethyl)sulfonyl]benzene
CAS Number:	1535-65-5
Molecular formula:	$C_7H_6F_2O_2S$
Molar mass:	192.18 g/mol
Appearance:	colourless solid

Density:	1.348 g/cm^3
Melting point:	24.7 – 25.0 °C
Boiling point:	115 - 120 °C at 0 hPa
Solubility in water:	soluble
Solubility:	soluble in acetonitrile

Preparation and properties:

Difluoromethyl phenyl sulfone derived from thiophenol according to below reaction scheme. Thiophenol react with difluorochloromethane, in various solvent using NaOH and tri[2-(2-methoxyethoxy)ethyl]amine to give difluoromethyl phenyl thioether,[1] which reacts with hydrogen peroxide in acetic acid gives difluoromethyl phenyl sulfone.[2]

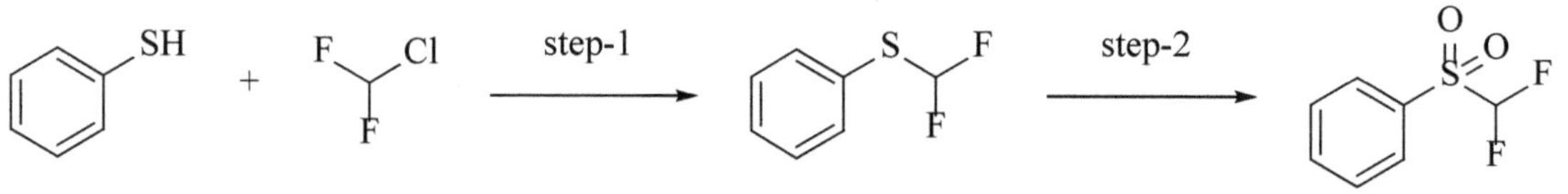

Scheme 1

Applications:

Building block for synthesis. Has been used for nucleophilic difluoro (phenyl-sulfonyl)methylation of carbonyls, reductive silylation and the preparation of trifluoroand difluoromethylsilanes, fluoroalkylation/ chloroalkylation of α,βenones, arynes, acetylenic ketones and other Michael acceptors and difluoromethylation of primary alkyl halides . It is also used in preparation of α-difluoromethyl amines, antidifluoropropanediols, β-difluoromethylated and β-difluoromethylenated alcohols and amines, difluoroalkenes, difluoromethyl alcohol derivatives and fluoromethylated vicinal ethylenediamines.

Reagent Used for

• Reductive silylation and the preparation of trifluoro- and difluoromethyl-silanes by reductive coupling of fluoromethyl sulfones, sulfoxides and sulfides with chlorosilanes [3]

• Fluoroalkylation/chloroalkylation of α,β-enones, arynes, acetylenic ketones and other Michael acceptors [4]

• Difluoromethylation of primary alkyl halides via nucleophilic substitution-reductive desulfonylation [5]

Reagent used in Preparation of

• α-difluoromethyl amines via stereoselective (phenylsulfonyl)difluoromethylation of chiral sulfinyl aldimines [6]

• Anti-difluoropropanediols via potassium tert-butoxide-catalyzed difluoromethylenation of aldehydes [7]

• β-difluoromethylated and β-difluoromethylenated alcohols and amines by regioselective nucleophilic difluoromethylation of 1,2-cyclic sulfates and sulfamidates [8]

• Difluoroalkenes from alkyl halides via nucleophilic substitution-elimination[9]

• Difluoromethyl alcohol derivatives from enolizable and non-enolizable carbonyl compounds using nucleophilic phenylsulfonyldifluoromethylation-reductive desulfonylation strategy [10]

• Fluoromethylated vicinal ethylenediamines via fluoromethylation of chiral α-aminobutanesulfinimines with (phenylsulfonyl)fluoromethanes followed by reductive desulfonylation and alcoholysis [11]

Safety:

Handle in accordance with good industrial hygiene and safety practice. Wash hands before breaks and at the end of workday. Handle with gloves. Gloves must be inspected prior to use. Impervious clothing, the type of protective equipment must be selected according to the concentration and amount of the dangerous substance at the specific workplace.

References:

1. *Journal of Fluorine chemistry.* **1988**, Vol.41, p. 247-262.

2. *Journal of American Chemical Society.* **1960**, Vol.82, p. 6178-6181.

3. G K Surya Prakash et. al. *Journal of Organic Chemistry.* **2003**, 68(11), 4457-4463.

4. Chuanfa Ni et. al. *Journal of Organic Chemistry.* **2008**, 73(15), 5699-5713.

5. G K Surya Prakash et. al. *Organic Letters.* **2004**, 6(23), 4315-4317.

6. Ya Li and Jinbo Hu. *Angewandte Chemie (International Edition).* **2005**, 44(36), 5882-5886.

7. G K Surya Prakash et. al. *Angewandte Chemie (International Edition).* **2003**, 42(42), 5216-5219.

8. Chuanfa Ni et. al. *Angewandte Chemie (International Edition).* **2007**, 46(5), 786-789.

9. G K Surya Prakash et. al. *Angewandte Chemie (International Edition).* **2004**, 43(39), 5203-5206.

10. G K Surya Prakash et. al. *European J. Org. Chem.* **2005**, 11, 2218-2223.

11. Jun Liu et. al. *Journal of Organic Chemistry.* 2007, 72(8), 3119-3121.

3.1.16. Fluoroboric acid

Fluoroboric acid or tetrafluoroboric acid is an inorganic compound with the chemical formula H_3OBF_4. It is mainly produced as a precursor to other fluoroborate salts.[1] It is a strong acid. Fluoroboric acid is corrosive and attacks the skin. It is available commercially as a solution in water and other solvents such as diethyl ether. With strength comparable to nitric acid, fluoroboric acid is a strong acid with a weakly coordinating, non-oxidizing conjugate base. It is structurally similar to perchloric acid, but lacks the hazards associated with oxidants.

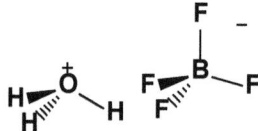

Structure:

IUPAC Name: Tetrafluoroboric acid

Other Name:	tetrafluoroboric acid, oxonium tetrafluoroboranuide
CAS Number:	(for H_3OBF_4), 16872-11-0 (solvent free), 80628-99-5 (for $H_5O_2BF_4$) 14219-41-1 (for H_3OBF_4), 16872-11-0 (solvent free), 80628-99-5 (for $H_5O_2BF_4$)
Molecular formula:	HBF_4
Molar mass:	87.81 g/mol
Appearance:	Colourless liquid
Density:	1.4 g/mL at 25 °C
Melting point:	−90 °C (−130 °F; 183 K)
Boiling point:	130 °C (266 °F; 403 K)
Solubility in water:	soluble in water

Structure and Production:

Although the solvent-free HBF_4 has not been isolated, its solvates are well characterized. These salts consist of protonated solvent as a cation, e.g., H_3O^+ and $H_5O_2^+$, and the tetrahedral BF_4^- anion. The anion and cations are strongly hydrogen-bonded.[2]

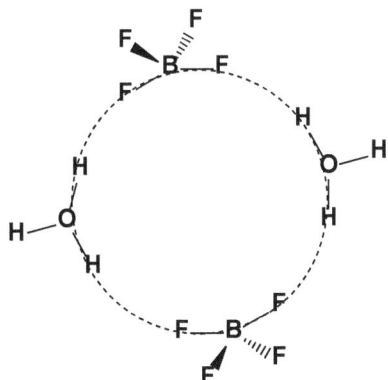

Subunit of crystal structure of H_3OBF_4 highlighting the hydrogen bonding between the cation and the anion.

Aqueous solutions of HBF_4 are produced by dissolving boric acid in aqueous hydrofluoric acid.[3][4] Three equivalents of HF react to give the intermediate boron trifluoride and the fourth gives fluoroboric acid:

$$B(OH)_3 + 4\,HF \rightarrow H_3O^+ + BF_4^- + 2\,H_2O$$

Anhydrous solutions can be prepared by treatment of aqueous fluoroboric acid with acetic anhydride.[5]

Applications:

Fluoroboric acid is the principal precursor to fluoroborate salts, which are typically prepared by treating the metal oxides with fluoroboric acid. The inorganic salts are intermediates in the manufacture of flame-retardant materials and glazing frits, and in electrolytic generation of boron. HBF_4 is also used in aluminum etching and acid pickling.

Organic chemistry

HBF_4 is used as a catalyst for alkylations and polymerizations. In carbohydrate protection reactions, ethereal fluoroboric acid is an efficient and cost-effective catalyst for transacetalation and isopropylidenation reactions. Acetonitrile solutions cleave acetals and some ethers, while neat fluoroboric acid removes tert-butoxycarbonyl groups. Many reactive cations have been obtained using fluoroboric acid, e.g. tropylium tetrafluoroborate ($C_7H_7^+BF_4^-$), triphenylmethyl tetrafluoroborate ($Ph_3C^+BF_4^-$), triethyloxonium tetrafluoroborate ($Et_3O^+BF_4^-$), and benzenediazonium tetrafluoroborate ($PhN_2^+BF_4^-$).

Galvanic cells

Aqueous HBF_4 is used as an electrolyte in galvanic cell oxygen sensor systems, which consist of an anode, cathode, and oxygen-permeable membrane. The solution of HBF_4 is able to dissolve lead(II) oxide from the anode in the form of lead tetrafluoroborate while leaving the rest of the system unchanged.

Metal plating

A mixture of CrO_3, HBF_4, and sulfonic acids in conjunction with a cathode treatment give tin-plated steel. Tin(I) fluoroborate/fluoroboric acid mixtures and organic reagents are used as the electrolyte in the cathode treatment of the tin plating process. Similar processes of electrodeposition and electrolytic stripping are used to obtain specific metal alloys.

A 10% solution of fluoroboric acid, tin fluoroborate and thiourea is used to electroless tin plate the copper traces of printed circuit boards for protection and easier soldering.

Safety:

Remove all contaminated clothes and footwear immediately unless stuck to skin. Drench the affected skin with running water for 10 minutes or longer if substance is still on skin. Transfer to hospital if there are burns or symptoms of poisoning. Wash out mouth with water. Do not induce vomiting. Give 1 cup of water to drink every 10 minutes. If unconscious, check for breathing and apply artificial respiration if necessary. If unconscious and breathing is OK, place in the recovery position.

References:

1. Gregory K. Friestad, Bruce P. Branchaud "Tetrafluoroboric Acid" *E-Eros Encyclopedia of Reagents for Organic Synthesis.* doi:10.1002/0470842-89X. rt035.

2. Mootz, D.; Steffen, M. "Crystal structures of acid hydrates and oxonium salts. XX. Oxonium tetrafluoroborates H_3OBF_4, $[H_5O_2]BF_4$, and $[H(CH_3OH)_2]BF_4$", *Zeitschrift für Anorganische und Allgemeine Chemie. 1981, vol. 482, pp. 193-200.* doi:10.1002/zaac.19814821124.

3. Brotherton, R. J.; Weber, C. J.; Guibert, C. R.; Little, J. L. (**2005**), "Boron Compounds", *Ullmann's Encyclopedia of Industrial Chemistry*, Weinheim: Wiley-VCH, doi:10.1002/14356007.a04_309.

4. Flood, D. T. "Fluorobenzene". *Org. Synth.* **1933,** 13: 46.; *Coll. Vol.* 2, p. 295.

5. Wudl, F.; Kaplan, M. L., "2,2′-Bi-L,3-Dithiolylidene (Tetrathiafulvalene, TTF) and its Radical Cation Salts" *Inorg. Synth.* **1979**, vol. 19, 27. doi:10.1002/ 9780470132500.

3.1.18. Fluolead

This compound has shown to promote nucleophilic deoxofluorination reactions and exhibits excellent thermal stability, up to 150 °C, therefore rendering it safely applicable in a variety of processes also in industrial scale production.

Structure:	
IUPAC Name:	4-tert-butyl-2, 6-dimethyl phenylsulfur trifluoride
Other Name:	Fluolead
CAS Number:	947725-04-4
Molecular formula:	$C_{12}H_{17}F_3S$
Molar mass:	250.32 g/mol
Appearance:	Off-white to pale brown solid
Density:	-
Melting point:	62-68 °C
Boiling point:	92-93 °C/0.5 mmHg
Solubility in water:	Reacts with water
Solubility:	very soluble in usual organic solvents.

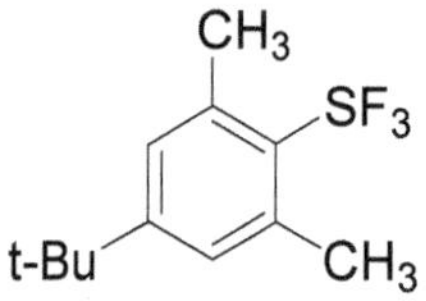

Preparation and properties:

Reaction Condition: Fluolead™ requires acid conditions, and hence the addition of a strong acid is preferable for enhancing the reactivity of the fluorination. In this respect, the addition of HF is most optimal for fluorinations with Fluolead, and the addition of HF/pyridine is particularly recommended.

Fluolead thermal stability: Fluolead has greater thermal stability than other deoxofluorinating agents such as DAST and Deoxo-fluor™. In differential scanning calorimetry (DSC) measurement of FLUOLEAD™, the thermal decomposition started at 260°C (Gold plated cell), and the released energy was 832 J/g, it is reported that DAST and Deoxo-fluor™ decomposed at 140°C (both are in Gold Cell), and the released energy was 1700 J/g and 1100 J/g, respectively).

Applications:

The reactivity and behaviour of Fluolead™ in fluorinations is very similar to SF4, however the reactivity and selectivity of our compound is generally superior to SF$_4$. Furthermore, as a solid, fluolead is easier and safer to handle. Some examples regarding to Fluolead are given below, alcohol convert in to fluoro, keto convert in to difluoro and acid convert in to trifluoride..[1]

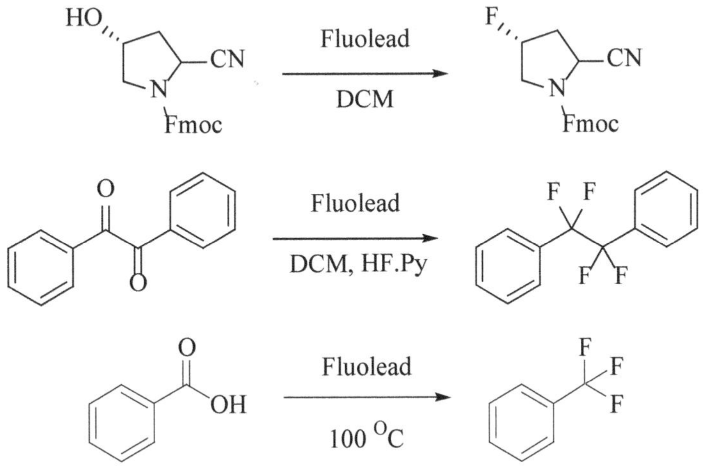

Scheme 1

Reacts with aldehyde gives difluoro compounds.[2]

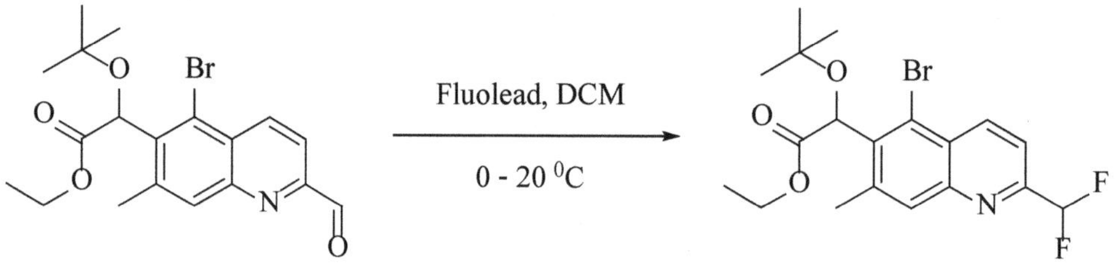

Scheme 2

Reacts with acid gives trifluoro compounds.[3]

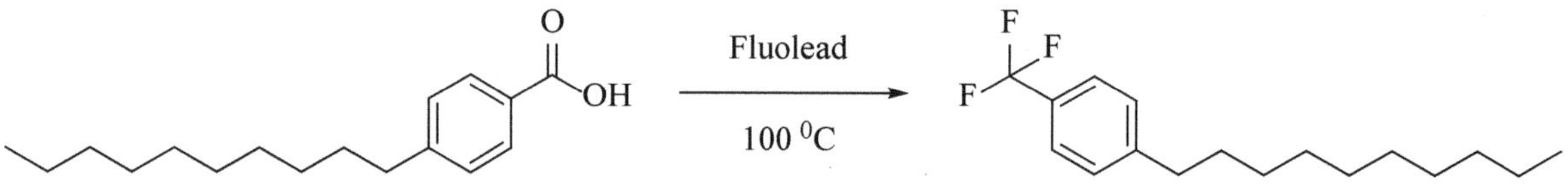

Scheme 3

Neutralization: Ar-SOF can be neutralized with alkaline solution, although this method can cause the formation of a variety of organosulfur side products. This is due to the disproportionation reaction(s) which can occur with sulfur compounds. Ar-SOF is NOT hydrolyzed with alkaline aqueous solution rapidly, thus the undesired side reactions (disproportionation) are allowed to occur simultaneously during the hydrolysis.

Alcohol Method: Our newly developed "alcohol method" appears to be most effective in terms of avoiding the formation of the unexpected disproportionation products. Ar-SOF reacts with alcohol (ROH) at room temperature, and forms the corresponding Ar-SO2R, Ar-Sulfinate, so that this method enables the removal of Ar-sulfinate as the "sole side product" from the reaction.

Safety:

Hydrofluoric (HF) acid burns require immediate and specialized first aid and medical treatment. Symptoms may be delayed up to 24 hours depending on the concentration of HF. After decontamination with water, further damage can occur due to penetration/ absorption of the fluoride ion. Treatment should be directed toward binding the fluoride ion as well as the effects of exposure. Skin exposures can be treated with a 2.5% calcium gluconate gel repeated until burning ceases.

More serious skin exposures may require subcutaneous calcium gluconate except for digital areas unless the physician is experienced in this technique, due to the potential for tissue injury from increased pressure. Absorption can readily occur through the subungual areas and should be considered when undergoing decontamination. Prevention of absorption of the fluoride ion in cases of ingestion can be obtained by giving milk, chewable calcium carbonate tablets or Milk of Magnesia to conscious victims. Conditions such as hypoc hypomagnesemia and cardiac arrhythmias should be monitored for, since they can occur after exposure. Consult a physician. Show this safety data sheet to the doctor in attendance.

References:

1. Umemoto, T.; Singh, R. P.; Xu, Y. *Journal of American Chemical Society.* **2010**, 132, 18199.

2. WO 2012/3498 A1, **2012**.

3. US 7265247 B1, **2007**.

3.1.18. Fluorosulfuric acid

Fluorosulfuric   acid (IUPAC name: sulfurofluoridic   acid)   is   the inorganic compound with the chemical formula HSO_3F. It is asuperacid and one of the strongest acids commercially available. The formula HSO_3F emphasizes its relationship to sulfuric acid, H_2SO_4; HSO_3F is a tetrahedral molecule. It is a colourless liquid although commercial samples are often yellow.[1]

Structure:	
IUPAC Name:	Fluorosulfuric acid
Other Name:	Fluorosulfonic acid, Fluorosulphonic acid, Fluorinesulphonic acid, Epoxysulfonyl fluoride
CAS Number:	7789-21-1
Molecular formula:	HFO_3S

Molar mass:	100.07 g/mol
Appearance:	Colourless liquid
Density:	1.84 g cm^{-3}
Melting point:	$-87.5 \text{ °C}; -125.4 \text{ °F}; 185.7 \text{ K}$
Boiling point:	$165.4 \text{ °C}; 329.6 \text{ °F}; 438.5 \text{ K}$
Solubility in water:	-

Preparation and properties:

Fluorosulfuric acid is prepared by the reaction of HF and sulfur trioxide: [1]

$$SO_3 + HF \rightarrow HSO_3F$$

Alternatively, KHF_2 or CaF_2 can be treated with oleum at 250 °C. Once freed from HF by sweeping with an inert gas, HSO_3F can be distilled in a glass apparatus.[2]

Chemical properties: Fluorosulfuric acid is a free-flowing colourless liquid. It is soluble in polar organic solvents (e.g. nitrobenzene, acetic acid, and ethyl acetate), but poorly soluble in nonpolar solvents such as alkanes. Reflecting its strong acidity, it dissolves almost all organic compounds that are even weak proton acceptors.[3] HSO_3F hydrolyzes slowly to HF and sulfuric acid. The related triflic acid (CF_3SO_3H) retains the high acidity of HSO_3F but is more hydrolytically stable. The self-ionization of fluorosulfonic acid also occurs:

$$2HSO_3F \rightleftharpoons [H_2SO_3F]^+ + [SO_3F]^- \quad K = 4.0 \times 10^{-8} \text{ (at 298K)}$$

HSO_3F is one of the strongest known simple Brønsted acids, although carborane-based acids are still stronger.[4] It has an H_0 value of -15.1 compared to -12 for sulfuric acid. The combination of HSO_3F and the Lewis acid antimony pentafluoride produces "Magic acid," which is a far stronger protonating agent. These acids all fall into the category of "superacids", acids stronger than 100% sulfuric acid.

Applications:

HSO$_3$F is useful for regenerating mixtures of HF and H$_2$SO$_4$ for etching lead glass. HSO$_3$F isomerizes alkanes and the alkylation of hydrocarbons with alkenes,[5] although it is unclear if such applications are of commercial importance. It can also be used as a laboratory fluorinating agent.[2]

Safety:

Fluorosulfuric acid is considered to be highly toxic and corrosive. It hydrolyzes to release HF. Addition of water to HSO$_3$F can be violent, similar to the addition of water to sulfuric acid but much more violent.

References:

1. Erhardt Tabel, Eberhard Zirngiebl, Joachim Maas "Fluorosulfuric Acid" in "Ullmann's Encyclopedia of Industrial Chemistry" **2005**, Wiley-VCH, Weinheim. doi:10.1002/14356007.a11_431.

2. Cotton, F. A.; Wilkinson, G. (1980). *Advanced Inorganic Chemistry* (4th ed.). New York: Wiley. p. 246. ISBN 0-471-02775-8.

3. Olah, G. A.; Prakash, G. K.; Wang, Q.; Li, X.-Y. (**2001**). "Fluorosulfuric Acid". *Encyclopedia of Reagents for Synthesis*. John Wiley & Sons.doi: 10.1002/047084289X.rf014.

4. Christopher A. Reed "Myths about the Proton. The Nature of H+ in Condensed Media" *Acc. Chem. Res.* **2013**, 46 (11), pp 2567–2575. doi:10.1021/ar 4000-64q.

5. Olah, G.; Farooq, O.; Husain, A.; Ding, N.; Trivedi, N.; Olah, J. "Superacid HSO$_3$F/HF-Catalyzed Butane Isomerisation". *Catalysis Letters. 1991*, 10 (3–4): 239–247. doi:10.1007/BF00772077.

3.1.19. Gold(III) fluoride

Gold(III) fluoride, AuF$_3$, is an orange solid that sublimes at 300 °C.[1] It is a powerful fluorinating agent.

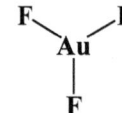

Structure:

IUPAC Name: Gold(III) fluoride

Other Name: Gold trifluoride, Auric fluoride

CAS Number: 14720-21-9

Molecular formula: AuF_3

Molar mass: 253.962 g/mol

Appearance: Orange-yellow **hexagonal** crystals

Density: 6.75 g/cm^3

Melting point: Sublimes above 300°C

Boiling point: -

Solubility in water: insoluble

Preparation and properties:

Gold (III) fluoride is available through the direct intersection between the elements in a monel metal container (to counter reactivity of fluorine gas)

$$2Au + 3F_2 \rightarrow AuF_3$$

Applications:

Gold (III) fluoride is used for the strengthen of gold caps on teeth.

References:

1. Greenwood, Norman N.; Earnshaw, Alan (**1997**). *Chemistry of the Elements* (2nd ed.). Butterworth-Heinemann. p. 1184. ISBN 0080379419.

2. F. W. B. Einstein, P. R. Rao, James Trotter and Neil Bartlett (**1967**). "The crystal structure of gold trifluoride". *Journal of the Chemical Society A: Inorganic, Physical, Theoretical* 4: 478–482. doi:10.1039/J19670000478.

3.1.20. Iodine pentafluoride

Iodine pentafluoride is an interhalogen compound with chemical formula IF$_5$. It is a fluoride of iodine.

Structure:

IUPAC Name: Iodine(V) fluoride

Other Name: Iodic fluoride

CAS Number: 7783-66-6

Molecular formula: IF$_5$

Molar mass: 221.89 g/mol

Appearance: Colourless or pale yellow liquid

Odour: irritating odour

Density: 3.250 g/cm^3

Melting point: 9.43 °C (48.97 °F; 282.58 K)

Boiling point: 97.85 °C (208.13 °F; 371.00 K)

Solubility in water: Reacts

Preparation and properties:

Iodine pentafluoride is a strong fluorination agent and is highly oxidative. It reacts vigorously with water forming hydrofluoric acid and with more fluorine forming iodine heptafluoride.

It was first synthesized by Henri Moissan in 1891 by burning solid iodine in fluorine gas.[1] This exothermic reaction is still used to produce iodine pentafluoride, although the reaction conditions have been improved.[2][3]

$$I_2 + 5\,F_2 \rightarrow 2\,IF_5$$

Iodine pentafluoride synthesized as per below method.[4]

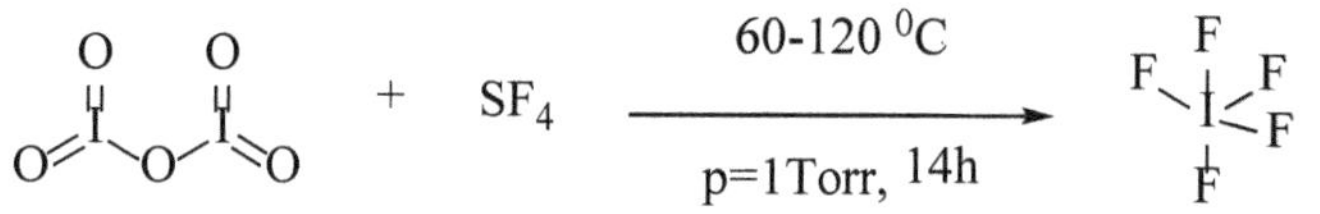

Applications:

Primary amines react with iodine pentafluoride forming nitriles after hydrolysis with water.[5]

$$R\text{-}CH_2\text{-}NH_2 \rightarrow R\text{-}CN$$

Safety:

Avoid contact with combustible materials. Do not touch spilled material. Stop leak if possible without personal risk. Reduce vapors with water spray. Do not get water inside container. Small spills: Flood with water. Large spills: Dike for later disposal. Keep unnecessary people away, isolate hazard area and deny entry. Ventilate closed spaces before entering.

References:

1. Moissan, M. H. "Nouvelles Recherches sur le Fluor". *Annales de Chimie et de Physique.* **1891**, 6 (24): 224–282.

2. Keim, R. (**1930**). "Das Jod-7-fluorid [The iodine-7-fluoride]". *Zeitschrift für Anorganische und Allgemeine Chemie* (in German). **1930**, 193(1): 176–186. doi:10. 1002/zaac.19301930117.

3. Ruff O.; Keim R. (**1931**). "Fluorierung von Verbindungen des Kohlenstoffs (Benzol und Tetrachlormethan mit Jod-5-fluorid, sowie Tetrachlormethan mit Fluor) [Fluoridation of Carbon Compounds (Benzene and Tetrachlormethane with Iodine-5-Fluoride, and Tetrachloromethane with Fluorine)]". *Zeitschrift für Anorganische und Allgemeine Chemie* (in German). **1931**, 201 (1): 245–258. doi:10.1002/zaac. 19312010122.

4. Patent US7132578 B$_1$, **2006**.

5. Stevens, T. E. "Rearrangement of Amides with Iodine Pentafluoride". *Journal of Organic Chemistry.* **1966,** 31 (6): 2025–2026.doi:10. 1021/jo01344a539.

3.1.21. Iodotoluene Difluoride

The importance of selectively fluorinating compounds in medicinal chemistry, biology, and organic synthesis is well appreciated and provides a major impetus to the discovery of new fluorinating agents that can operate according to an efficient, safe, and mild criteria. Elemental fluorine, and many electrophilic fluorinating agents have been used in synthesis; however, most of these fluorinating agents are highly aggressive, unstable, and require special equipment and care for safe handling. By contrast, iodotoluene difluoride (tol-IF$_2$) is easy to handle, and is less toxic than many fluorinating agents. Sigma-Aldrich is pleased to introduce this new reagent for oxidation and fluorination.

Structure:

IUPAC Name: 4-Iodotoluene difluoride

Other Name: P-Tolyliododifluoride

CAS Number: 371-11-9

Molecular formula: C$_7$H$_7$F$_2$I

Molar mass: 256.03 g/mol

Appearance: Solid

Density: -

Melting point: 102-106 °C

Boiling point: -

Solubility in water: -

Solubility: very soluble in usual organic solvents

Preparation and properties:

4-iodotoluene reacts with Fluorine in monofluoro trichloromethane to yield 4-Iodotoluene difluoride. [1]

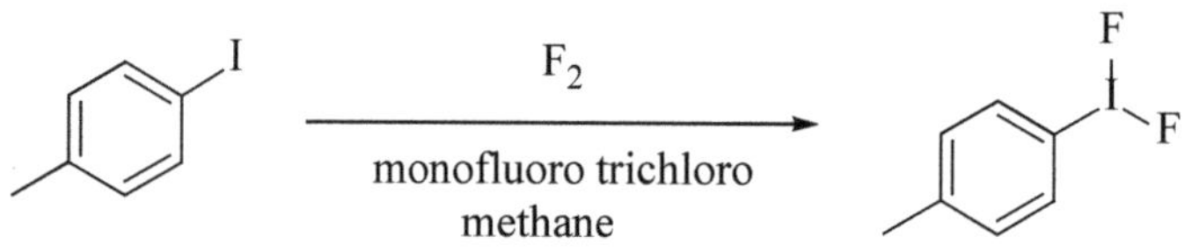

Scheme 1

In second route of synthesis 4-iodotoluene dichloride reacts with HF in the presence of mercury oxide to give 4-Iodotoluene difluoride.[2]

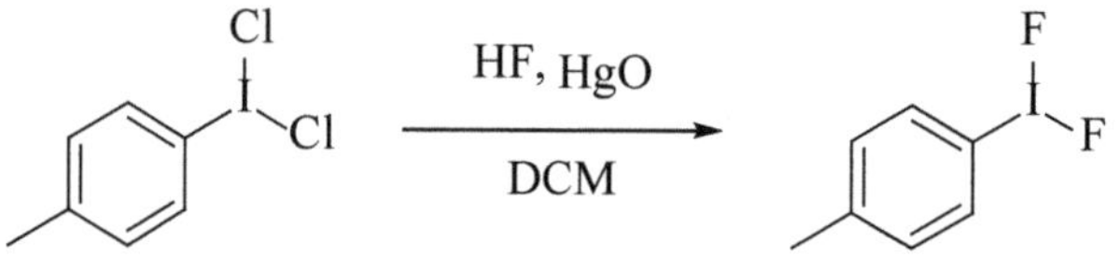

Scheme 2

Applications:

A new methodology for the synthesis of fluorinated cyclic ethers was recently reported, which utilized tol-IF$_2$ to achieve a fluorinative ring-expansion of four-, five-, and six-membered rings, an example of which is illustrated below (Scheme 3).[3]

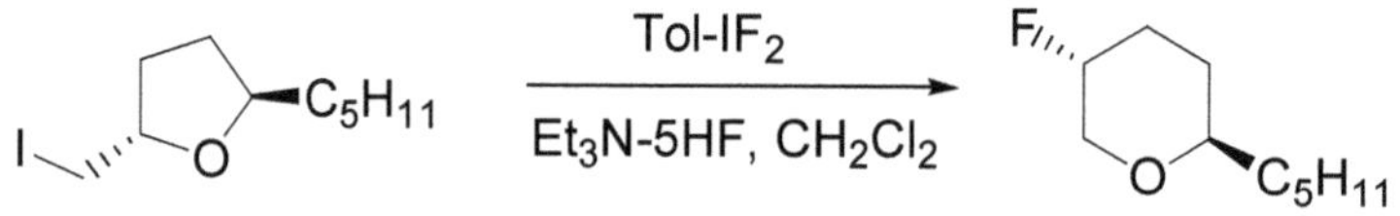

Scheme 3

Selective monofluorination of β-ketoesters, β-ketoamides, and diketones takes place without requirement of HF-amine complexes under mild conditions (Scheme 4). The formation of difluoro products were not detected in these reactions.[4]

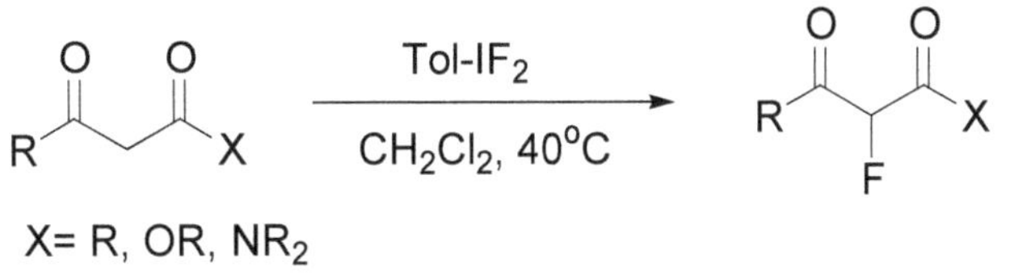

Scheme 4

When one equivalent of tol-IF$_2$ is reacted with phenylsulfanylated esters, the α-fluoro sulfide results through a Fluoro Pummerer reaction. In contrast to diethylaminosulfur trifluoride, two equivalents produces the α,α-difluoro sulfide, and three equivalents of tol-IF$_2$ produces the α,α-difluoro sulfoxide. This behaviour was exploited in the one pot synthesis of a 3-fluoro-2(5H)-furanone (Scheme 5).[5]

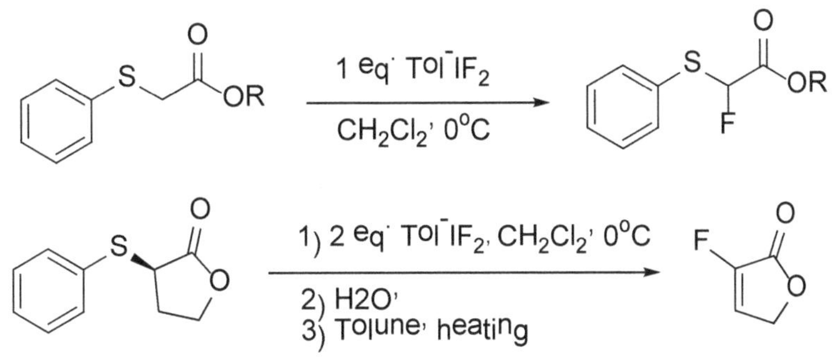

Scheme 5

Treatment of phenylsulfanylated lactams with one equivalent of tol-IF$_2$ results in the unsaturated heterocycle in moderate to good yields. When two equivalents of the reagent were used, the lactams were fluorinated in the α - and β-positions resulting in the diastereomeric difluoride (Scheme 6).[6]

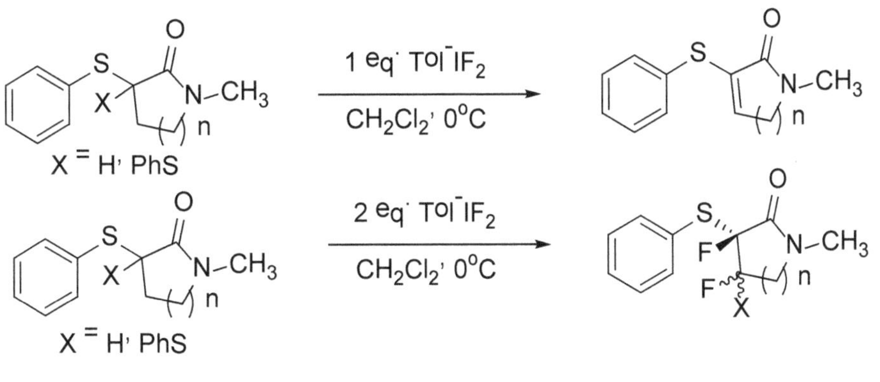

Scheme 6

Safety:

Wash contaminated clothing before reuse. Discard contaminated shoes. Wash thoroughly after handling. Use respirators and components tested and approved under appropriate government standards such as NIOSH (US) or CEN (EU). Where risk assessment shows air-purifying respirators are appropriate use a full-face particle respirator type N100 (US) or type P3 (EN 143) respirator cartridges as a backup to engineering controls.

If inhaled, remove to fresh air. If not breathing give artificial respiration. If breathing is difficult, give oxygen. In case of contact with eyes, flush with copious amounts of water for at least 15 minutes. Assure adequate flushing by separating the eyelids with fingers. Call a physician. Emits toxic fumes under fire conditions.

References:

1. *Journal of Fluorine Chemistry.* **1980**, Vol.15, p. 213-222.
2. *Journal of Fluorine Chemistry.* **2000**, Vol.104, # 2, p. 277-280.
3. Inagaki, T. et al. *Tetrahedron Lett.*, **2003**, *44*, 4117.
4. Yoshida, M. et al. *ARKIVOC* **2003**, (vi) 36.
5. Motherwell, W. B. et al. *J. Chem. Soc. Perkin Trans.* 1, **2002**, 2809.
6. Greaney, M. F. et al. *Tetrahedron Lett.* **2001**, *42*, 8523.

3.1.22. Ishikawa's reagent

Ishikawa's reagent is a fluorinating reagent used in organic chemistry. It is used to convert alcohols into alkyl fluorides and carboxylic acids into acyl fluorides. Aldehydes and ketones do not react with it. The reagent consists of a mixture of *N*,*N*-diethyl-(1,1,2,3,3,3-hexafluoropropyl)amine and *N*,*N*-diethyl-(*E*)-pentafluoropropenylamine in varying proportions. The active species is the hexafluoropropylamine; any enamine is converted into this by the hydrogen fluoride by-product as the reaction proceeds.

Ishikawa's reagent is a popular alternative to the DAST reagent, since it is shelf-stable and easily prepared from inexpensive and innocuous reagents. It is an improvement on Yarovenko's reagent, the adduct of chlorotrifluoro-ethylene and diethylamine, which must be prepared in a sealed vessel and once prepared keeps only for a few days, even in the refrigerator.

The reagent is mostly used to convert primary alcohols to alkyl fluorides under mild conditions with high yield. However, secondary and tertiary alcohols give a substantial amount of alkenes and ethers as side products.

Structure:

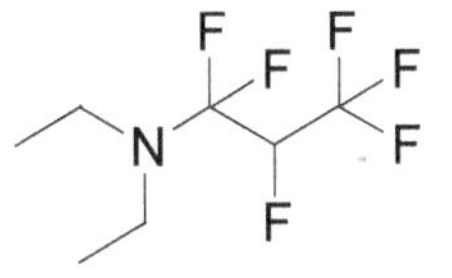

IUPAC Name:	*N,N*-Diethyl-1,1,2,3,3,3-hexafluoropropylamine
Other Name:	Ishikawa's reagent
CAS Number:	309-28-6
Molecular formula:	$C_7H_{11}F_6N$
Molar mass:	223.16 g/mol
Appearance:	Yellow liquid
Density:	1.230 g/mL at 25 °C(lit.)
Melting point:	-
Boiling point:	56-57 °C(lit.)
Solubility in water:	-
Solubility:	Organic solvents

Preparation and properties:

The compound is prepared by adding hexafluoropropene to a solution of diethyl amine in ether at 0 °C and distilling the product *in vacuo*. The amount of enamine in the product depends on temperature control during the reaction – the higher the temperature the more enamine.[1]

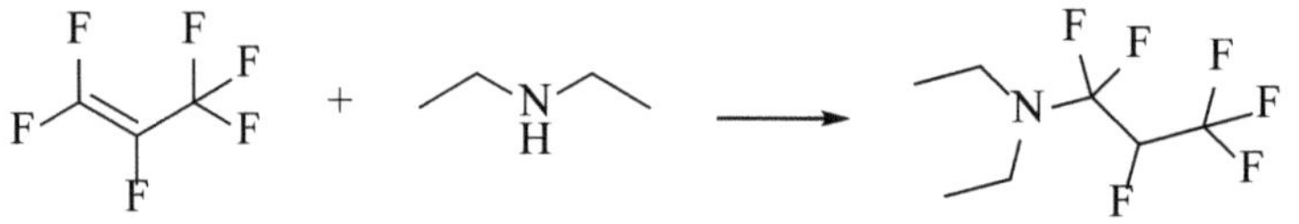

Scheme 1

Applications:

Ishikawa's reagent some applications are as per below.[2][3]

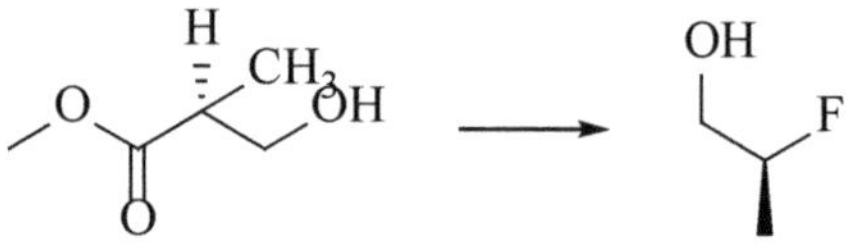

Scheme 2

Ishikawa's reagent also displays the ability to insert a fluoro in unsaturated alcohols.

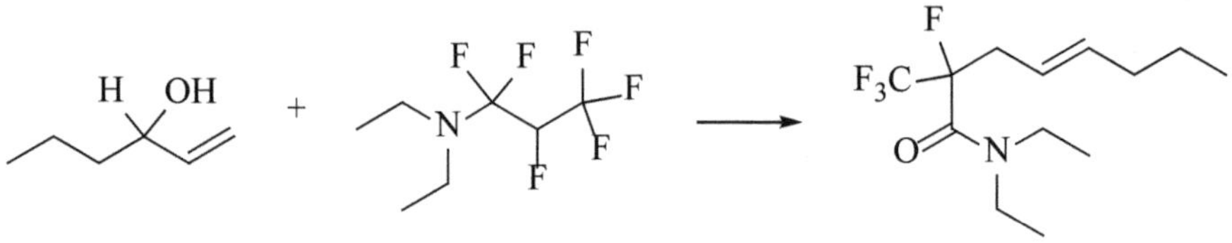

Scheme 3

Perfluorinated alkylamines, such as Ishikawa's reagent (N,N-Diethyl-1,1,2,3,3,3-hexafluoropropylamine),[4] are highly selective for hydroxyl groups and do not react with aldehydes and ketones. The amide byproducts of these reagents, however, are harder to separate from the desired products than aminosulfurane byproducts.

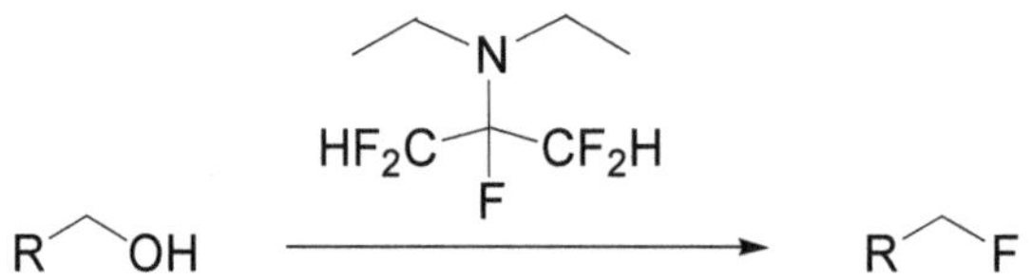

Scheme 4

Safety:

Use personal protective equipment. Avoid breathing vapours, mist or gas. Ensure adequate ventilation. Remove all sources of ignition. Evacuate personnel to safe areas. Beware of vapours accumulating to form explosive concentrations. Vapours can accumulate in low areas.

References:

1. *Journal of American Chemical Society.* **1960**, Vol. 82, p. 5116-5122.

2. Patent US2014/357661A1, **2014**.

3. *Journal of Fluorine chemistry.* **2003**, Vol. 124, # 1, p. 69-80.

4. Takaoka, A.; Iwagiri, H.; Ishikawa, N. *Bull. Chem. Soc. Jpn.* **1979**, Vol. *52*, p. 3377.

3.1.23. Manganese(III) fluoride

Manganese(III) fluoride (also known as Manganese trifluoride) is the inorganic compound with the formula MnF_3. This red/purplish solid is useful for converting hydrocarbons into fluorocarbons, i.e., it is a fluorination agent.[1] It also forms a hydrate.

Structure:

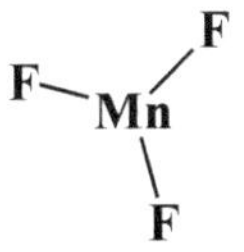

IUPAC Name: Manganese(III) fluoride

Other Name: Manganese trifluoride, manganic fluoride

CAS Number: 7783-53-1

Molecular formula: MnF_3

Molar mass: 111.938 g/mol

Appearance: Purple-pink powder

Density: 3.54 g/cm^3

Melting point: > 600 °C (1,112 °F; 873 K) (decomposes)

Boiling point: -

Solubility in water: Hydrolysis

Preparation and properties:

MnF_3 can be prepared by treating a solution of MnF_2 in hydrogen fluoride with fluorine:[2]

$MnF_2 + 0.5\ F_2 \rightarrow MnF_3$

It can also be prepared by the reaction of elemental fluorine with a manganese (II) halide at ~250 °C.[3]

MnF_3 reacts with sodium fluoride to give the octahedral hexafluorate anion:[3]

$3NaF + MnF_3 \rightarrow Na_3MnF_6$

Other reaction conditions give compounds with anion formula MnF_5^{2-} or MnF_4^{-}. These anions are chain and layer structures respectively, with bridging fluorine. Manganese remains 6 coordinate, octahedral, and trivalent in all these materials.[3]

Manganese(III) fluoride fluorinates organic compounds including aromatic hydrocarbons,[4] cyclobutenes,[5] and fullerenes.[6]

On heating, MnF_3 decomposes to manganese(II)fluoride.[7][8]

Applications:

Starting material for solid state preparation of Na_3MnF_6 for further high pressure structural determination studies.[9]

Structure:

In the crystalline state, MnF_3 resembles vanadium(III) fluoride: both feature octahedral metal centers with the same average M-F bond distances. In the Mn compound, however, is distorted (and hence a monoclinic unit cell vs. a higher symmetry one) due to the Jahn-Teller effect, with pairs of Mn-F distances of 1.79, 1.91, 2.09 Å.[10][11][12]

The hydrate $MnF_3.3H_2O$ is obtained by crystallisation of MnF_3 from hydrofluoric acid. The hydrate is unusual in that it forms two different structures (both based on $[Mn(H_2O)_4F_2]+\ [Mn(H_2O)_2F_4]-$), which have space groups P21/c and P21/a.[13]

Relate Mn(II) compounds

Other manganese(III) compounds include manganese(III) acetate (CAS# 993-02-2), manganese acetylacetonate (CAS# 14284-89-0), Both are employed as oxidants in organic synthesis. MnF_3 is Lewis acidic and forms a variety of derivatives. Two examples are $K_2MnF_3(SO_4)$[14] and K_2MnF_5.

Safety:

Like other reactive inorganic fluorides, MnF_3 should be stored in a polyethylene bottle and contact with skin or any other moist area avoided due to the formation of Hydrofluoric acid on hydrolysis.

References:

1. Burley, G. A.; Taylor, R. "Manganese(III) fluoride" in Encyclopedia of Reagents for Organic Synthesis (Ed: L. Paquette) **2004**, *J. Wiley & Sons, New York*. doi:10.1002/047084289.

2. Z. Mazej. "Room temperature syntheses of MnF_3, MnF_4 and hexafluoro-manganete(IV) salts of alkali cations". *Journal of Fluorine Chemistry.* **2002**, 114 (1): 75–80. doi:10.1016/S0022-1139(01)00566-8.

3. Inorganic chemistry, Catherine E. Housecroft, A.G. Sharpe, pp.711-712, section *Manganese (III)*, googlebooks link.

4. Fluorination of p-chlorobenzotrifluoride by manganese trifluoride A. Kachanov, V. Kornilov, V.Belogay, Fluorine Notes :Vol. 1 (1), **1998**, via*notes. fluorine1.ru.*

5. Fluorination of fluoro-cyclobutene with high-valency metal fluoride m Junji Mizukado, Yasuhisa Matsukawa, Heng-dao Quan, Masanori Tamura, Akira Sekiya, *Journal of Fluorine Chemistry.* Vol. 127, Issue 1, **2006**, P.79-84, online abstract via*www.sciencedirect.com.*

6. Fluorination of the cubic and hexagonal C60 modifications by crystalline manganese trifluoride , Physics of the Solid State . Vol. 44, **2002** , pp.629-

630 , V.É. Aleshina, A.Ya. Borshchevskii, E.V. Skokan, I.V. Arkhangel'skii, A.V. Astakhov, N.B. Shustova , online abstract via *www.springerlink.*

7. Manganese; section *Manganic Salts* via *www.1911encyclopedia.org.*

8. In situ time-resolved X-ray diffraction study of manganese trifluoride thermal decomposition , J.V. Raua, V. Rossi Albertinib, N.S. Chilingarova, S. Colonnab, U. Anselmi Tamburini, *Journal of Fluorine Chemistry.* 4506 (**2001**) 1–4 , online version.

9. Carlson, S. et al. *Inorg. Chem.* 37, 1486, (**1998**).

10. Wells, A.F. (**1984**) Structural Inorganic Chemistry, Oxford: Clarendon Press. ISBN 0-19-855370-6.

11. Hepworth, M. A.; Jack, K. H.; Nyholm, R. S. (**1957**). "Interatomic Bonding in Manganese Trifluoride". *Nature.* 179 (4552): 211–212. doi:10. 1038/179211b0.

12. M. A. Hepworth, K. H. Jack (**1957**). "The crystal structure of manganese trifluoride, MnF$_3$". *Acta Crystallographica.* 10 (5): 345–351. doi:10.1107/ S0365110X57001024.

13. Structures of two polymorphs of MnF$_3$·3H$_2$O , Michel Molinier and Werner Massa , Journal of Fluorine Chemistry , Vol. 57, Issues 1-3, April–June **1992**, p.139-146 ,online abstract via *www.sciencedirect.com.*

14. Bhattacharjee, M. N; Chaudhuri, M. K. (**1990**). "Dipotassium Trifluoro-sulfatomanganate(III)". *Inorg. Synth.* 27: 312–313. doi:10. 1002/9780470132 586.

3.1.24. Manganese(IV) tetrafluoride

Manganese tetrafluoride, MnF$_4$, is the highest fluoride of manganese. It is a powerful oxidizing agent and is used as a means of purifying elemental fluorine.[1][2]

Structure:

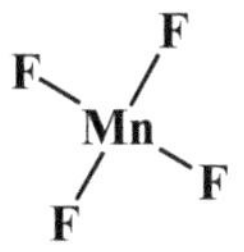

IUPAC Name:	manganese tetrafluoride
Other Name:	manganese(IV) fluoride
CAS Number:	15195-58-1
Molecular formula:	MnF_4
Molar mass:	130.93 g/mol
Appearance:	Blue solid
Density:	3.61 g cm^{-1} (calc.)
Melting point:	70 °C (158 °F; 343 K) *decomposes*
Boiling point:	Decomposes
Solubility in water:	Reacts violently

Preparation and properties:

Manganese tetrafluoride was first unequivocally prepared in 1961 by the reaction of manganese(II) fluoride (or other Mn^{II} compounds) with a stream of fluorine gas at 550 °C: the MnF_4 sublimes into the gas stream and condenses onto a cold finger.[3][4] This is still the commonest method of preparation, although the sublimation can be avoided by operating at increased fluorine pressure (4.5–6 bar at 180–320 °C) and mechanically agitating the powder to avoid sintering of the grains.[1][5] The reaction can also be carried out starting from manganese powder in a fluidized bed.[6][7]

Other preparations of MnF_4 include the fluorination of MnF_2 with krypton difluoride,[8] or with F_2 in liquid hydrogen fluoride solution under ultraviolet light.[9] Manganese tetrafluoride has also been prepared (but not isolated) in an acid–base reaction between antimony pentafluoride and K_2MnF_6 as part of a chemical synthesis of elemental fluorine.[10]

$$K_2MnF_6 + 2\ SbF_5 \rightarrow MnF_4 + 2\ KSbF_6$$

Manganese tetrafluoride reacts violently with water and even with sodium-dried petroleum ether. It immediately decomposes on contact with moist air.[3]

Reaction with alkali metal fluorides or concentrated hydrofluoric acid gives the yellow hexafluoromanganate(IV) anion $[MnF_6]^{2-}$.[11]

Decomposition:- Manganese tetrafluoride is in equilibrium with manganese (III) fluoride and elemental fluorine:

$$MnF_4 \rightleftharpoons MnF_3 + FRACTION\ F_2$$

Decomposition is favoured by increasing temperature, and disfavoured by the presence of fluorine gas, but the exact parameters of the equilibrium are unclear, with some sources saying that MnF_4 will decompose slowly at room temperature,[12][13] others placing a practical lower temperature limit of 70 °C,[1][14] and another claiming that MnF_4 is essentially stable up to 320 °C.[11] The equilibrium pressure of fluorine above MnF_4 at room temperature has been estimated at about 10^{-4} Pa (10^{-9} bar), and the enthalpy change of reaction at +44(8) kJ mol^{-1}.[15]

Applications:

The main application of manganese tetrafluoride is in the purification of elemental fluorine. Fluorine gas is produced by electrolysis of anhydrous hydrogen fluoride (with a small amount of potassium fluoride added as a support electrolyte) in a Moissan cell. The technical product is contaminated with HF, much of which can be removed by passing the gas over solid KF, but also with oxygen (from traces of water) and possibly heavy-metal fluorides such as arsenic pentafluoride (from contamination of the HF). These contaminants are particularly problematic for the semiconductor industry, which uses high-purity fluorine for etching silicon wafers. Further impurities, such as iron, nickel, gallium and tungsten compounds, can be introduced if unreacted fluorine is recycled.[2]

The technical-grade fluorine is purified by reacting it with MnF_3 to form manganese tetrafluoride. As this stage, and heavy metals present will form in volatile complex fluorides, while the HF and O_2 are unreactive. Once the MnF3 has been converted, the excess gas is vented for recycling, carrying the remaining gaseous impurities with it. The MnF_4 is then heated to 380 °C to release fluorine at purities of up to 99.95%, reforming MnF_3, which can be reused.[1][2] By placing two reactors in parallel, the purification process can be made continuous, with one reactor taking in technical fluorine while the other delivers high-grade fluorine.[2] Alternatively, the manganese tetrafluoride can be isolated and transported to where the fluorine is needed, at lower cost and greater safety than pressurized fluorine gas.[1][5]

Fluoromanganate(IV) complexes

The yellow hexafluoromanganate (2−) of alkali metal and alkaline earth metal cations have been known since 1899, and can be prepared by the fluorination of MnF_2 in the presence of the fluoride of the appropriate cation.[9][16][17][18] They are much more stable than manganese tetrafluoride.[10] Potassium hexafluoromanganate(IV), K_2MnF_6, can also be prepared by the controlled reduction of potassium permanganate in 50% aqueous hydrofluoric acid.[19][20]

$$2\ KMnO_4 + 2\ KF + 10\ HF + 3\ H_2O_2 \rightarrow 2\ K_2MnF_6 + 8\ H_2O + 3\ O_2$$

The pentafluoromanganate (1−) salts of potassium, rubidium and caesium, $MMnF_5$, can be prepared by fluorination of $MMnF_3$ or by the reaction of $[MnF_4(py)(H_2O)]$ with MF.[18][20] The lemon-yellow heptafluoromanganate(3−) salts of the same metals, M_3MnF_7, have also been prepared.[21]

Safety:

Hazards and risks associated with manganese: manganese metal powder is a fire hazard. Unless known otherwise, all manganese compounds should be regarded as highly toxic as well as possibly carcinogenic and teratogenic.

References:

1. WO, "Method of manufacturing manganese tetrafluoride", published 2006-03-30.

2. WO, "Process for the purification of elemental fluorine", published 2009-06-18.

3. Hoppe, Rudolf; Dähne, Wolfgang; Klemm, Wilhelm (**1961**), "Mangan-tetrafluorid, MnF_4", *Naturwissenschaften* 48 (11): 429. doi:10.1007/ BF-00621676.

4. Hoppe, Rudolf; Dähne, Wolfgang; Klemm, Wilhelm (**1962**), "Mangantetrafluorid mit einem Anhang über $LiMnF_5$ und $LiMnF_4$", *Justus Liebigs Ann. Chem.* 658 (1): 1–5. doi:10.1002/jlac. 19626580102.

5. WO, "Method for preparing manganese tetrafluoride", published 2009-06-18.

6. Roesky, H.; Glemser, O. (**1963**), "A New Preparation of Manganese Tetrafluoride",*Angew. Chem., Int. Ed. Engl.* 2 (10): 626. doi:10.1002/ anie.196306262.

7. Roesky, Herbert W.; Glemser, Oskar; Hellberg, Karl-Heinz (**1965**), "Darstellung von Metallfluoriden in der Wirbelschicht", *Chem. Ber.* 98 (6): 2046–48. doi:10.1002/cber.19650980642.

8. Lutar, Karel; Jesih, Adolf; Žemva, Boris (**1988**), "KrF_2/MnF_4 adducts from KrF_2/MnF_2 interaction in HF as a route to high purity MnF_4". *Polyhedron* 7 (13): 1217–19. doi:10.1016/S02775387(00)812127.

9. Mazej, Z. (**2002**), "Room temperature syntheses of MnF_3, MnF_4 and hexa-fluoromanganete(IV) salts of alkali cations". *J. Fluorine Chem.* 114 (1): 75–80. doi:10.1016/S0022-1139(01)00566-8.

10. Christe, Karl O. (**1986**), "Chemical synthesis of elemental fluorine". *Inorg. Chem.* 25(21): 3721–24. doi:10.1021/ic00241a001.

11. Adelhelm, M.; Jacob, E. (**1991**), "MnF_4: preparation and properties". *J. Fluorine Chem.* 54 (1–3): 21. doi:10.1016/S0022-1139(00)83531-9.

12. Cotton, F. Albert; Wilkinson, Geoffrey (**1980**). *Advanced Inorganic Chemistry* (4th ed.), New York: Wiley, p. 745. ISBN 0-471-02775-8.

13. Housecroft, Catherine E.; Sharpe, Alan G. (**2007**). *Inorganic Chemistry* (3rd ed.), New York: Prentice Hall, p. 710. ISBN 0131755536.

14. Rakov, E. G.; Khaustov, S. V.; Pomadchin, S. A. (**1997**), "Thermal Decompo-sition and Pyrohydrolysis of Manganese Tetrafluoride". *Russ. J. Inorg. Chem.* 42 (11): 1646–4.

15. Ehlert, T. C.; Hsia, M. (**1972**), "Mass spectrometric and thermochemical studies of the manganese fluorides". *J. Fluorine Chem.* 2 (1): 33–51. doi: 10.1016/S0022-1139(00)83113-9.

16. Weinland, R. F.; Lauenstein, O. (**1899**). *Z. Anorg. Allg. Chem.* 20: 40.

17. Hoppe, Rudolf; Blinne, Klaus (**1957**), "Hexafluoromanganate IV der Elemente Ba, Sr, Ca und Mg". *Z. Anorg. Allg. Chem.* 291 (5–6): 269–75. doi:10.1002/ zaac.19572910507.

18. Hoppe, Rudolf; Liebe, Werner; Dähne, Wolfgang (**1961**), "Über Fluoro-manganate der Alkalimetalle". *Z. Anorg. Allg. Chem.* 307 (5–6): 276–89. doi: 10.1002/zaac.19613070507.

19. Bode, Hans; Jenssen, H.; Bandte, F. (**1953**), "Über eine neue Darstellung des Kalium hexafluoromanganats(IV)". *Angew. Chem.* 65 (11): 304. doi: 10.1002/ ange.19530651108.

20. Chaudhuri, M. K.; Das, J. C.; Dasgupta, H. S. (1981), "Reactions of KMnO$_4$ A novel method of preparation of pentafluoromanganate (IV)[MnF$_5$]$^-$". *J. Inorg. Nucl. Chem.* 43 (1): 85–87. doi:10.1016/00221-902(81)80440-X.

21. Hofmann, B.; Hoppe, R. (**1979**), "Zur Kenntnis des (NH$_4$)$_3$SiF$_7$-Typs. Neue Metallfluoride A$_3$MF$_7$ mit M = Si, Ti, Cr, Mn, Ni und A = Rb, Cs". *Z. Anorg. Allg. Chem.* 458 (1): 151–62. doi:10.1002/zaac.19794580-121.

3.1.25. Mercury(II) fluoride

Mercury(II) fluoride has the molecular formula Hg F$_2$.

Structure:

IUPAC Name: Mercury(II) fluoride

Other Name:	Mercuric fluoride
CAS Number:	7783-39-3
Molecular formula:	F_2Hg
Molar mass:	238.59 g/mol
Appearance:	White cubic crystals
Density:	8.95 g/cm³
Melting point:	645 °C (decomposed)
Solubility in water:	reacts

Preparation and properties:

Mercury(II) fluoride is most commonly produced by the reaction of mercury(II) oxide and hydrogen fluoride:

$$HgO + 2\ HF \rightarrow HgF_2 + H_2O$$

Mercury(II) fluoride can also be produced through the fluorination of mercury(II) chloride:

$$HgCl_2 + F_2 \rightarrow HgF_2 + Cl_2$$

or mercury(II) oxide:.[1]

$$2\ HgO + 2\ F_2 \rightarrow 2\ HgF_2 + O_2$$

Applications:

Mercury(II) fluoride is a selective fluorination agent.[2]

Safety:

Avoid contact with skin, eyes and clothing. Wash hands before breaks and immediately after handling the product.

Handle with gloves. Gloves must be inspected prior to use. Use proper glove removal technique (without touching glove's outer surface) to avoid skin contact with this product. Dispose of contaminated gloves after use in accordance with applicable laws and good laboratory practices.

Complete suit protecting against chemicals, the type of protective equipment must be selected according to the concentration and amount of the dangerous substance at the specific workplace.

References:

1. Greenwood, Norman N.; Earnshaw, Alan (**1997**), *Chemistry of the Elements* (2nd ed.). Butterworth-Heinemann. ISBN 0080379419.

2. Habibi, Mohammed H.; Mallouk, Thomas E. (**1991**), "Photochemical selective fluorination of organic molecules using mercury (II) fluoride". *Journal of Fluorine Chemistry* **51** (2): 291. doi:10.1016/S00221139(00) 80299-7.

3.1.26. Morpholinosulfur trifluoride

Structure:	
IUPAC Name:	4-(Trifluoro-λ^4-sulfanyl)morpholine
Other Name:	Morpho-DAST, Morpholinosulfar, Morpholinosulfur trifluoride
CAS Number:	51010-74-3
Molecular formula:	$C_4H_8F_3NOS$
Molar mass:	175.17 g/mol
Appearance:	clear yellow to amber liquid
Odour:	Irritating
Density:	1.436 g/mL at 25 °C(lit.)

Melting point: -

Boiling point: 41-42 °C/0.5 mm Hg(lit.)

Solubility in water: Reacts

Solubility: Organic nonpolar solvents

Preparation and properties:

Morpho-DAST is synthesized according to below reaction scheme. First morpholine react with disulfarchloride in hexane using sodium hydroxide as base to yield dimer and finally reacts with chlorine in presence of potassium fluoride to yield morpho-DAST.[1][2]

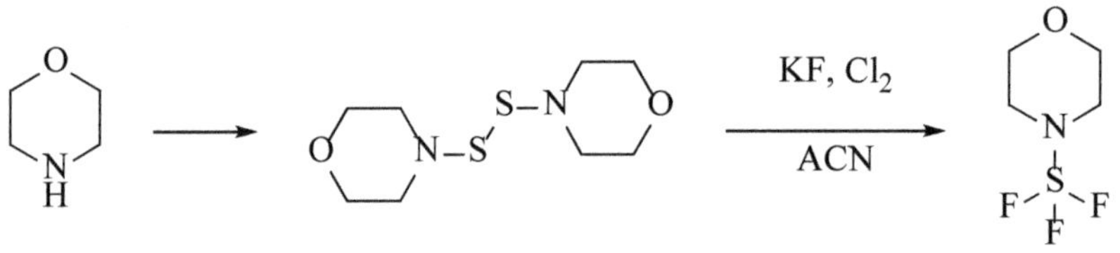

Scheme 1

Applications:

A nucleophilic fluorinating agent. Some examples are given below. Hydroxyl conver to Fluoride,[3] carboxylic acid to acid fluoride.[4]

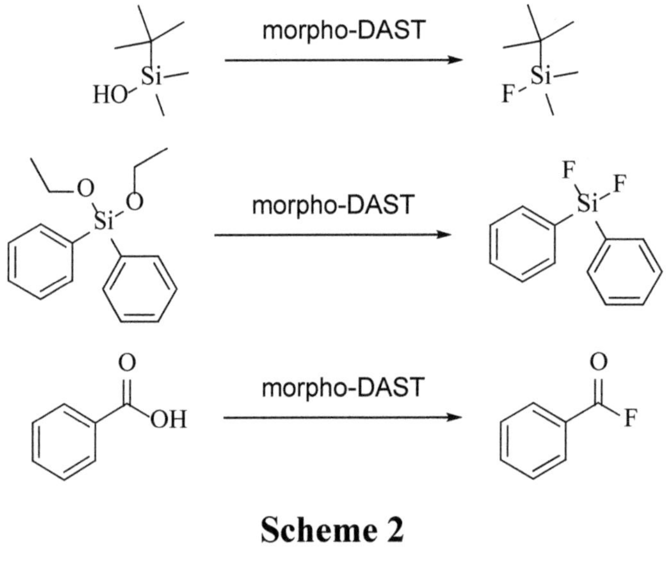

Scheme 2

Morpho-DAST reaction with aldehyde gives difluoro compound.[5]

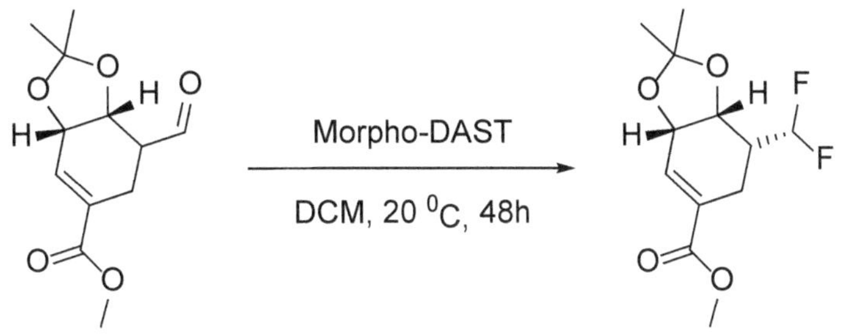

Scheme 3

Safety:

Self reactive substances, Causes severe skin burns and eye damage, Heating may cause an explosion, Wear protective gloves/ protective clothing/ eye protection/ face protection, In case of contact with eyes, rinse immediately with plenty of water and seek medical advice.

Skin protection

Handle with gloves. Gloves must be inspected prior to use. Use proper glove removal technique (without touching glove's outer surface) to avoid skin contact with this product. Dispose of contaminated gloves after use in accordance with applicable laws and good laboratory practices. Wash and dry hands.

Body Protection

Complete suit protecting against chemicals, Flame retardant protective clothing, the type of protective equipment must be selected according to the concentration and amount of the dangerous substance at the specific workplace.

Respiratory protection

Where risk assessment shows air-purifying respirators are appropriate use a full-face respirator with multi-purpose combination (US) or type ABEK (EN 14387) respirator cartridges as a backup to engineering controls. If the respirator is the sole means of protection, use a full-face supplied air respirator. Use respirators and components tested and approved under appropriate government standards such as NIOSH (US) or CEN (EU).

References:

1. Tetrahedron. Vol.47, # 20, p.3353-3364.

2. Synthetic communications. **2003**, Vol.33, # 14, p.2505-2509.

3. Patent US6344579 B$_1$, **2002**.

4. Patent US6329529 B$_1$, **2001**.

5. Tetrahedron. **1992**, vol. 48, # 40, p.8751-8774.

3.1.27. N-Fluorobenzenesulfonimide

Structure:

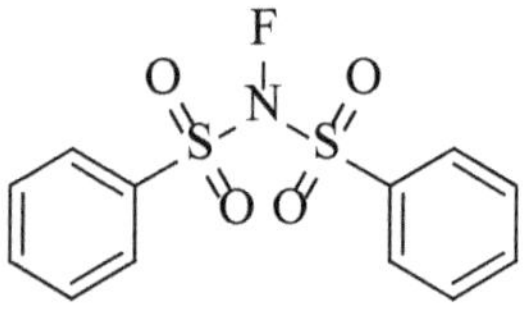

IUPAC Name: N-Fluorobenzenesulfonamide

Other Name: NFSI, N-Fluorobenzenesulphonimide

CAS Number: 133745-75-2

Molecular formula: $C_{12}H_{10}FNO_4S_2$

Molar mass: 315.34 g/mol

Appearance: White crystalline solid

Density: -

Melting point: 114 - 116 °C (decomposed)

Solubility in water: -

Solubility: DMF, DMAc organic solvent.

Preparation and properties:

One of the most convenient routes to prepare NFSI in high yield is the fluorination with 10% v/v F_2 in N_2 of N-(Phenylsulfonyl)benzenesulfonamide in MeCN at -40 °C.[1]

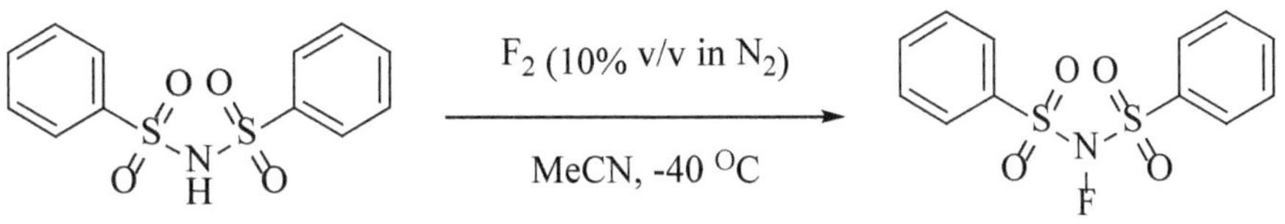

Scheme 1

N-Fluorobenzenesulfonamide also synthesized from the sodium salt of N-(Phenylsulfonyl)benzenesulfonamide.[2]

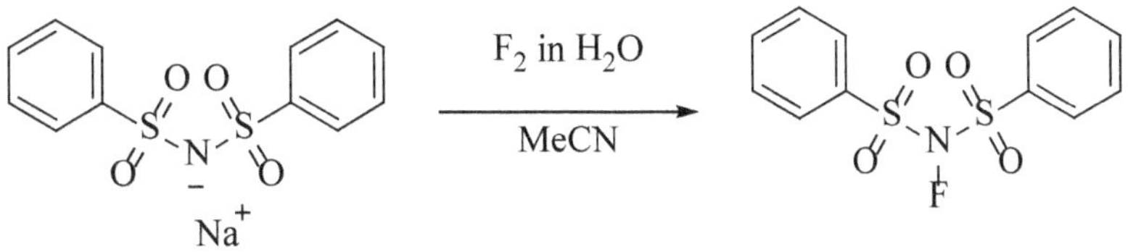

Scheme 2

Applications:

Diethyl (1-chloro-2-(4-fluorophenyl)-2-oxoethyl)phosphonate reacts with NFSI gives diethyl (R)-(1-chloro-1-fluoro-2-(4-fluorophenyl)-2 -oxoethyl) phosphornate.[3]

Scheme 3

Tert-butyl 2-phenoxyacetate reacts with NFSI gives (fluoromethoxy)benzene.[4]

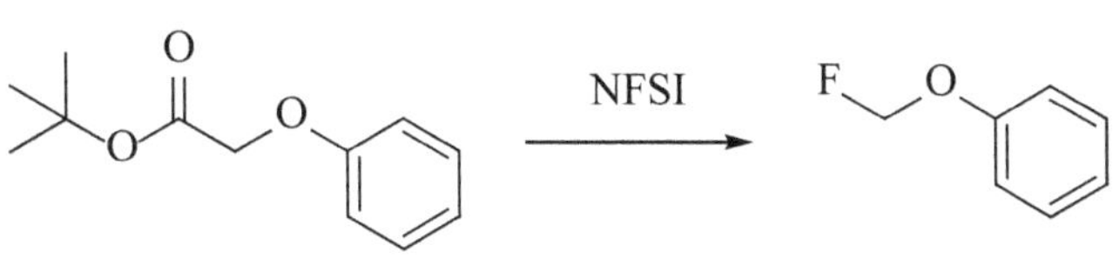

Scheme 4

Methyl-2,2-diphenylacetate reacts with NFSI gives Methyl-2-fluoro-2,2-diphenyl-acetate.[5]

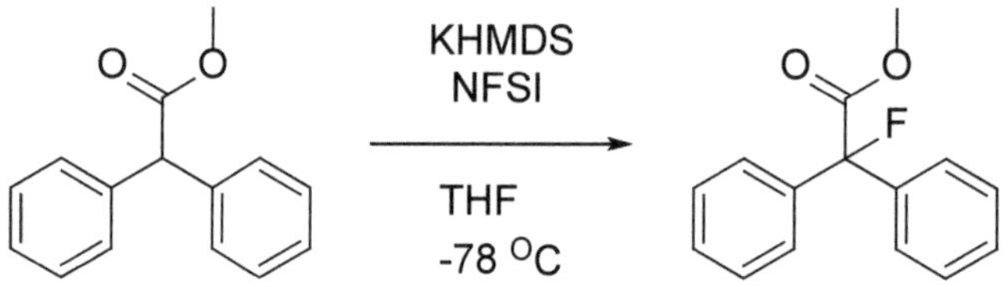

Scheme 5

Safety:

Handle with gloves. Gloves must be inspected prior to use. Use proper glove removal technique (without touching glove's outer surface) to avoid skin contact with this product. Dispose of contaminated gloves after use in accordance with applicable laws and good laboratory practices. Wash and dry hands. The type of protective equipment must be selected according to the concentration and amount of the dangerous substance at the specific workplace.

No component of this product present at levels greater than or equal to 0.1% is identified as probable, possible or confirmed human carcinogen by IARC.

References:

1. *Tetrahedron.* **2013**, Vol. 69, # 24, p. 4933 – 4937.
2. *Chemical Communications.* **2007**, # 23, p. 2330 – 2332.
3. *Tetrahedron Letters.* **2013**, Vol. 54, # 26, p. 3359 – 3362.
4. *Journal of American Chemical Society.* **2012**, Vol. 134, # 9, p. 4026 – 4029.
5. *Synlett.* **1991**, 187.

3.1.28. Nitrosyl fluoride

Nitrosyl fluoride, NOF, is a covalently bonded nitrosyl compound.

Structure:

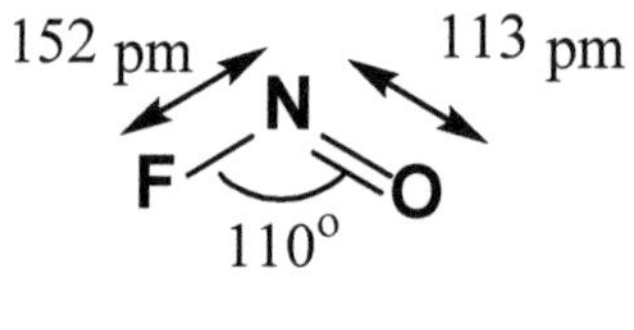

IUPAC Name: Nitrosyl fluoride

Other Name: Nitrosyl fluoride

CAS Number: **7789-25-5**

Molecular formula: NOF

Molar mass: 49.0045 g/mol

Appearance: Colourless gas

Density: 2.657 mg/mL

Melting point: −166 °C (−267 °F; 107 K)

Boiling point: −72.4 °C (−98.3 °F; 200.8 K)

Solubility in water: Reacts

Preparation and properties:

NOF synthesized from nitrosyl chloride.[1]

$$HF \quad + \quad O{\small\nwarrow}N{\nearrow}Cl \quad \longrightarrow \quad O{\small\nwarrow}N{\nearrow}F$$

NOF also synthesized from the reaction of nitrosonym ion and oxygenedifluoride.[2]

$$F{\small\diagdown}O{\diagup}F \quad + \quad {}^{-}N{=}O \quad \longrightarrow \quad \underset{F}{N}{\overset{O}{=}}$$

NOF is a highly reactive fluorinating agent that converts many metals to their fluorides, releasing nitric oxide:

$$n \, NOF + M \rightarrow MF_n + n \, NO$$

NOF also fluorinates fluorides to form adducts that have a salt-like character, such as $NOBF_4$. Aqueous solutions of NOF are powerful solvents for metals, by a mechanism similar to that seen in aqua regia. Nitrosyl fluoride reacts with water to form nitrous acid, which then forms nitric acid:

$$NOF + H_2O \rightarrow HNO_2 + HF$$

$$3 \, HNO_2 \rightarrow HNO_3 + 2 \, NO + H_2O$$

Nitrosyl fluoride can also convert alcohols to nitrites:

$$ROH + NOF \rightarrow RONO + HF$$

It has a bent molecular shape.

Applications:

Nitrosyl fluoride is used as a solvent and as a fluorinating and nitrating agent in organic synthesis. It has also been proposed as an oxidizer in rocket propellants.

Safety:

Wash thoroughly after handling. Minimize dust generation and accumulation. Do not breathe dust, mist, or vapour. Do not get in eyes, on skin, or on clothing. Keep container tightly closed. Do not ingest or inhale. Do not allow contact with water. Use only in a chemical fume hood. Keep from contact with moist air and steam.

References:

1. *Angewandte chemie.* **1961**, Vol. 73, p. 531-532.

2. *Z. Anorg. Chem.* **1931**, Vol. 198, p. 39-52.

3.1.29. Nitryl fluoride

Nitryl fluoride, NO_2F, is a colourless gas and strong oxidizing agent, which is used as a fluorinating agent[1] and has been proposed as an oxidiser in rocket propellants (though never flown). It is a molecular species, not ionic, consistent with its low boiling point. The structure features planar nitrogen with a short N-F bond length of 135 pm.[2]

Structure:	
IUPAC Name:	Fluoro(oxo)Azane Oxide
Other Name:	Fluorine Nitrite; Nitrogen Oxyfluoride
CAS Number:	10022-50-1

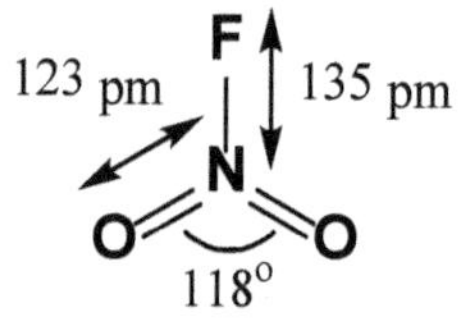

Molecular formula:	NO_2F
Molar mass:	65.0039 g/mol
Appearance:	-
Density:	1.39 g/cm^3
Melting point:	−166 °C (−267 °F; 107 K)
Boiling point:	−72 °C (−98 °F; 201 K)
Solubility in water:	-

Preparation and properties:

Henri Moissan and Lebeau recorded the preparation of nitryl fluoride in 1905 by the fluorination of nitrogen dioxide. This reaction is highly exothermic, which leads to contaminated products. The simplest method avoids fluorine gas but uses cobalt(III) fluoride:[3]

$$NO_2 + CoF_3 \rightarrow NO_2F + CoF_2$$

The CoF_2 can be regenerated to CoF_3. Other methods have been described.[4]

Themodynamic property:

The thermodynamic properties of this gas were determined by IR and Raman spectroscopy[5] The standard heat of formation of FNO_2 is -19 ± 2 kcal/mol.3

- The equilibrium of the unimolecular decomposition of FNO_2 lies on the side of the reactants by at least six orders of magnitude at 500 kelvin, and two orders of magnitude at 1000 kelvin.[5]

- The homogeneous thermal decomposition cannot be studied at temperatures below 1200 kelvin.[5]

- The equilibrium shifts towards the reactants with increasing temperature.[5]

- The dissociation energy of 46.0 kcal of the N-F bond in nitryl fluoride is about 18 kcal less than the normal N-F single bond energy. This can be attributed to the "reorganization energy" of the NO_2 radical; that is, the

NO_2 radical in FNO_2 is less stable than the free NO_2 molecule. Qualitatively speaking, the odd electron "used up" in the N-F bond forms a resonating three-electron bond in free NO_2, thus stabilizing the molecule with a gain of 18 kcal.[5]

Nitryl fluoride can be used to prepare organic nitro compounds and nitrate esters.

Applications:

Nitryl fluoride has been used as an oxidizer in rocket propellants and as a fluorinating agent.

Safety:

Poison by inhalation. A severe irritant to skin, eyes, and mucous membranes. A powerful oxidizing agent. This gas is intensely reactive. Explosive reaction with hydrogen at 200–300°C. Ignites on contact with antimony, arsenic, boron, iodine, phosphorus, selenium. Ignites when warmed with bismuth, carbon, chromium, lead, sulfur. Incandescent reaction with aluminum, cadmium, cobalt, iron, molybdenum, nickel, potassium, sodium, thorium, titanium, tungsten, uranium, vanadium, zinc, zirconium, lithium (at 200–300°C), manganese (at 200–300°C). Incompatible with metals, nonmetals. When heated to decomposition it emits toxic fumes of F^- and NO_x.

References:

1. Merck Index, 13th edition (**2001**), p.1193.

2. F.A.Cotton and G.Wilkinson. *Advanced Inorganic Chemistry*, 5th edition (**1988**), Wiley, p.333.

3. Davis, Ralph A.; Rausch, Douglas A. "Preparation of Nitryl Fluoride". *Inorganic Chemistry* 2 (6): 1300–1301. doi:10.1021/ic50010-a048.

4. Faloon, Albert V.; Kenna, William B. *Journal of the American Chemical Society* 73 (6): 2937–2938. doi:10.1021/ja01150a505.

5. Tschuikow-Roux, E. "THERMODYNAMIC PROPERTIES OF NITRYL FLUORIDE". *Journal of Physical Chemistry* 66 (9): 1636–1639. doi:10.1021/j100815a017.

3.1.30. Perchloryl fluoride

Perchloryl fluoride[1] is a reactive gas with the chemical formula ClO3F. It has a characteristic sweet odour[2] that resembles gasoline and kerosene. It is toxic and is a powerful oxidizing and fluorinating agent. It is the acid fluoride of perchloric acid.

In spite of its small enthalpy of formation ($\Delta H_f^{\ominus} = -5.7$), it is stable, decomposing only at 400 °C.[3] It is quite reactive towards reducing agents and anions, however, with the chlorine atom acting as an electrophile.[3] It reacts explosively with reducing agents such as amides, metals, hydrides, etc.[2] Its hydrolysis in water occurs very slowly, unlike that of chloryl fluoride.

Structure:

IUPAC Name: Perchloryl fluoride

Other Name: Chlorine oxyfluoride, Perchlorofluoride, Chlorine fluorine oxide, Trioxychlorofluoride, Perchloric acid fluoride

CAS Number: 7616-94-6

Molecular formula: ClO_3F

Molar mass: 102.4496 g/mol

Appearance: Colourless gas

Odour: Sweet odour

Density: 1.4 g/cm^3

Melting point: −147.8 °C (−234.0 °F; 125.3 K)

Boiling point:	−46.7 °C (−52.1 °F; 226.5 K)
Solubility in water:	0.06 g/100 ml (20 °C)
Viscocity:	3.91 x 10^{-3} Pa.s (@ melting point)

Preparation and properties:

Perchloryl fluoride is produced primarily by the fluorination of perchlorates. Antimony pentafluoride is a commonly used fluorinating agent:[3]

$$ClO_4^- + 3\ HF + 2\ SbF_5 \rightarrow ClO_3F + H_3O+ + 2\ SbF_6^- ClO_3F$$

reacts with alcohols to produce alkyl perchlorates, which are extremely shock-sensitive explosives.[4] Using Friedel-Crafts catalysts, it can be used for introducing the $-ClO_3$ group into aromatic rings via electrophilic aromatic substitution.[5]

Applications:

Perchloryl fluoride is used in organic chemistry as a mild fluorinating agent.[1] It was the first industrially relevant electrophilicfluorinating agent, used since the 1960s for producing fluorinated steroids.[4]

Perchloryl fluoride was investigated as a high performance liquid rocket fuel oxidizer.[6] In comparison with chlorine pentafluoride andbromine pentafluoride, it has significantly lower specific impulse, but does not tend to corrode tanks. It does not require cryogenic storage.

It can also be used in flame photometry as an excitation source.[7]

Safety:

Perchloryl fluoride is toxic, with a TLV of 3 ppm.[8] It is a strong lung- and eye-irritant capable of producing burns on exposed skin. Its IDLH level is 100 ppm.[9] Symptoms of exposure include dizziness, headaches, syncope, and cyanosis. Exposure to toxic levels causes severe respiratory tract inflammation and pulmonary edema.[6]

References:

1. Chemical Science and Technology Laboratory. "Perchloryl fluoride". National Institute of Standards and Technology. Retrieved 2009-11-28.

2. Jared Ledgard (**2007**). *The Preparatory Manual of Explosives* (3rd ed.). Lulu. com. p. 77. ISBN 0-615-14290-7.

3. Harry Julius Emeléus; A. G. Sharpe (**1976**). *Advances in inorganic chemistry and radiochemistry, Volume 18*. Academic Press. ISBN 0-12-023618-4.

4. Peer Kirsch (**2004**). *Modern fluoroorganic chemistry: synthesis, reactivity, applications*. Wiley-VCH. p. 74. ISBN 3-527-30691-9.

5. Peter Bernard David De la Mare (**1976**). *Electrophilic halogenation: reaction pathways involving attack by electrophilic halogens on unsaturated compounds*. CUP Archive. p. 63. ISBN 0-521-29014-7.

6. John Burke Sullivan; Gary R. Krieger (**2001**). *Clinical environmental health and toxic exposures* (2nd ed.). Lippincott Williams & Wilkins. p. 969. ISBN 0-683-08027-X.

7. Schmauch, G. E.; Serfass, E. J. (**1958**). "The Use of Perchloryl Fluoride in Flame Photometry". *Applied Spectroscopy* **12** (3): 98–102. Bibcode: 1958ApSpe..12...98S. doi:10.1366/000370258774615483.

8. National Institute for Occupational Safety and Health. "NIOSH Pocket Guide to Chemical Hazards". Centers for Disease Control and Prevention. Retrieved 2013-10-31.

9. National Institute for Occupational Safety and Health. "Documentation for Immediately Dangerous To Life or Health Concentrations (IDLHs)". Centers for Disease Control and Prevention. Retrieved 2013-10-31.

3.1.31. Perfluorobutanesulfonyl fluoride

Perfluorobutanesulfonyl fluoride (nonafluorobutanesulfonyl fluoride, NfF) is a colourless, volatile liquid that is immiscible with water but soluble in common organic solvents. It is prepared by the electrochemical fluorination of sulfolane. NfF serves as an entry point to nonafluorobutanesulfonates (nonaflates), which are valuable as electrophiles in palladium catalyzed cross coupling reactions. As a perfluoroalkylsulfonylating agent, NfF offers the advantages of lesser cost and greater stability over the more frequently used triflic anhydride. The fluoride leaving group is readily substituted by nucleophiles such as amines, phenoxides, and enolates, giving sulfonamides, aryl nonaflates, and alkenyl nonaflates respectively. However, it is not attacked by water (in which it is stable at pH<12). Hydrolysis by barium hydroxide gives Ba(ONf)$_2$, which upon treatment with sulfuric acid gives perfluorobutanesulfonic acid and insoluble barium sulfate.

Structure:

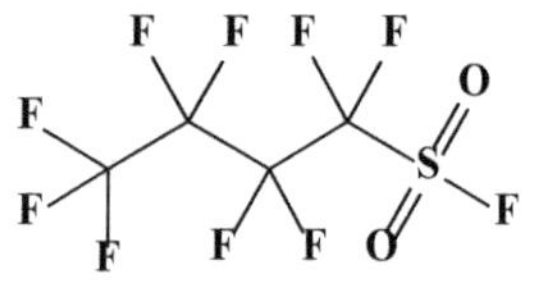

IUPAC Name: 1,1,2,2,3,3,4,4,4-nonafluorobutane-1-sulfonyl fluoride

Other Name: NfF

CAS Number: 375-72-4

Molecular formula: $C_4F_{10}O_2S$

Molar mass: 302.09 g/mol

Appearance: -

Density: 1.682 g/mol

Melting point: $< -120\,°C$ ($-184\,°F$; 153 K)

Boiling point: 65 to 66 °C (149 to 151 °F; 338 to 339 K)

Solubility in water: reacts

Preparation and properties:

Reaction of Perfluorobutanesulfonyl chloride with ammonium fluoride in the presence of HF give Perfluorobutanesulfonyl fluoride.[1]

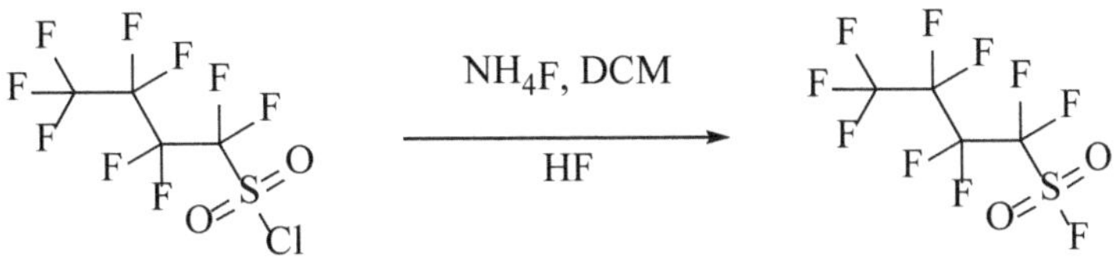

Scheme 1

Tetrhydrothiophene-1,1-dioxide reacts with HF gives Perfluorobutanesulfonyl fluoride.[2]

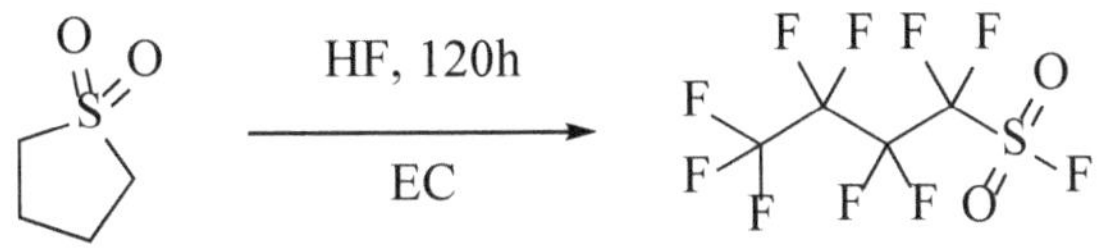

Scheme 2

Commercially available NfF is contaminated with 6-10 mol % perfluoro-sulfolane derived from its production. This is readily removed by vigorously stirring the commercial material with a concentrated aqueous solution of K_3PO_4 and K_2HPO_4 in a 1:1 molar ratio for 96 hours. This treatment, followed by removal of the aqueous layer and distillation from P_2O_5, gives a product that contains >99 mol % NfF with near quantitative recovery.[3]

Applications:

Synthesis of aryl and alkenyl nonaflates

As mentioned above, aryl and alkenyl nonaflates are useful as electrophiles in palladium catalyzed cross coupling reactions. Their reactivity generally mirrors that of the more commonly encountered triflate electrophiles, but nonaflates tend to be less prone to hydrolysis to ketones (in the case of alkenyl sulfonates) and phenols (in the case of aryl sulfonates). Their resistance to hydrolysis makes nonaflates superior electrophiles in Buchwald-Hartwig couplings, where this side reaction can be deleterious to yields of the desired product.[4]

The sodium enolates of β-ketoesters react with 1.15 equivalents of NfF to give the corresponding alkenyl nonaflates in high yield. Ethyl 2-methylacetoacetate (R=R'=Me) gives the geometrically pure E isomer by this method.[5]

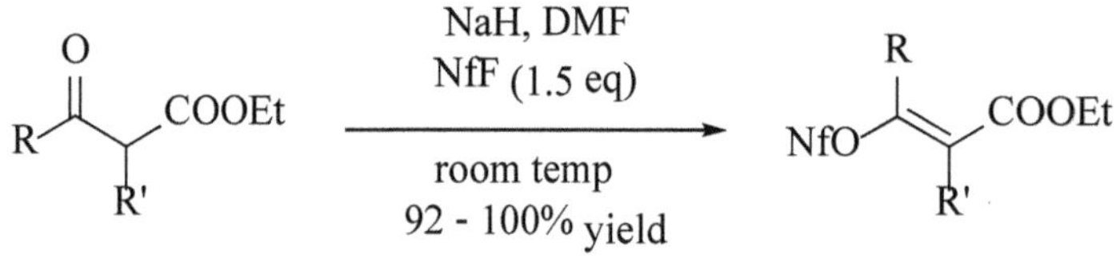

Scheme 3

Simple aldehydes and ketones react with NfF in the presence of bases such as DBU or phosphazenes to give alkenyl nonaflates in high yields without formation of a discrete enolate. Use of the P_2 phosphazene base at -30 to -20°C gives the less substituted alkenyl nonaflate with unsymmetrically substituted ketones.[3] Similar reactions with triflic anhydride generally require the use of the expensive 2,6-di-*tert*-butylpyridine to achieve high yields.

The reaction of enolates with NfF depends strongly both on the structure of the enolate and its metal counterion. The lithium enolates of methyl ketones give mixtures of products derived from electrophilic attack on the O (expected) or C (unexpected) atoms of the enolate. This effect is particularly evident with the lithium enolate of pinacolone, which gives a 2:1 mixture favoring C-attack. More substituted lithium enolates give only products of O sulfonylation in variable yields.[6]

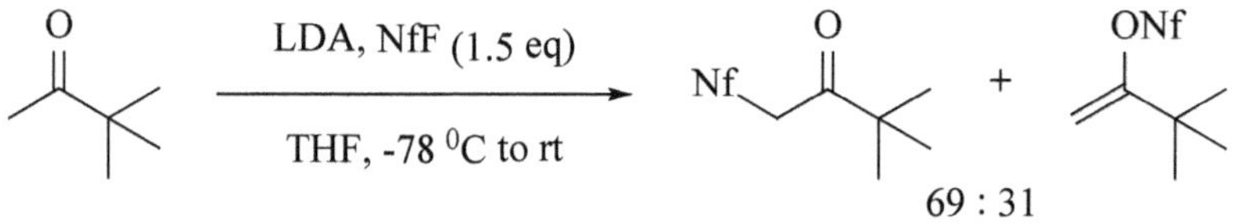

Scheme 4

Trimethylsilyl enol ethers react with NfF in the presence of a substoichiometric fluoride source at 0°C to ambient temperature to give alkenyl nonaflates in moderate to good yields. Dried Bu_4F was the preferred fluoride source in one study,[6] but CsF has been used in difficult cases with excellent results.[7]

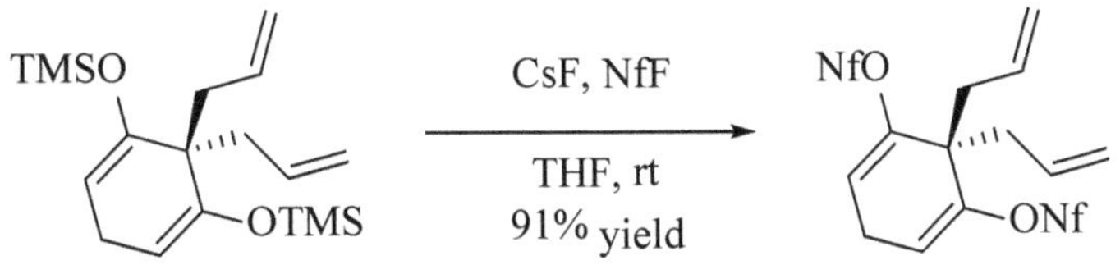

Scheme 5

Aryl nonaflates can be prepared straightforwardly from phenols and NfF in the presence of bases such as potassium carbonate[8] and Et$_3$N[4] in near quantitative yields. Stronger bases such as NaH and BuLi[9] can also be used, but they tend to give somewhat lower yields.

Reaction With Alcohols

The reaction of NfF with alcohols highlights the lability of alkyl nonaflates – in most cases, the final product of the reaction is either an alkyl fluoride (from F$^-$ attack on the intermediate alkyl nonaflate) or an olefin (from elimination of NfOH from the intermediate nonaflate).

Synthesis of *bis*-nonafluorobutanesulfonimide (Nf$_2$NH)

NfF reacts with ammonium chloride in the presence of triethylamine in acetonitrile to give the triethylammonium salt of the superacidic *bis*-nonafluorobutanesulfonimide in 97% yield. The corresponding potassium salt is obtained by treatment of a methanolic solution of the triethylammonium salt with KOH.[10] The acid is obtained by ion exchange chromatography of the triethylammonium salt with Amberlite IR-100 as the stationary phase and methanol as the eluent.[11] The actual species produced in the latter procedure is likely MeOH$_2^+$ Nf$_2$N$^-$.

Safety:

May be harmful if inhaled. Material is extremely destructive to the tissue of the mucous membranes and upper respiratory tract. Avoid breathing of airborne material. Select one of the following NIOSH approved respirators based on airborne concentration of contaminants and in accordance with OSHA regulations: Half-mask organic vapor respirator with dust/mist prefilter. This product contains one or more organic fluorochemicals that have the potential to be absorbed and remain in the body for long periods of time, either as the parent molecule or as metabolites, and may accumulate with repeated exposures. There are no known human health.

References:

1. *Journal of Fluorine chemistry.* **2007**, Vol. 128, # 11, p. 1353-1358.

2. *Journal of Fluorine chemistry.* **2004**, Vol. 125, # 2, p. 243-252.

3. Vogel, Michael A. K.; Christian B. W. Stark; Ilya M. Lyapkalo (2007). "A Straightforward Synthesis of Alkenyl Nonaflates from Carbonyl Compounds Using Nonafluorobutane-1-sulfonyl Fluoride in Combination with Phosphazene Bases". *Synlett* **2007** (EFirst): 2907–2911. doi:10. 1055/s-2007-991084. ISSN 0936-5214.

4. Meadows, Rebecca E.; Simon Woodward (2008-02-11). "Steric effects in palladium-catalysed amination of aryl triflates and nonaflates with the primary amines PhCH(R)NH$_2$ (R=H, Me)". *Tetrahedron* 64 (7): 1218–1224. doi:10. 1016/j.tet.2007.11.074. ISSN 0040-4020.

5. Bellina, Fabio; Donatella Ciucci; Renzo Rossi; Piergiorgio Vergamini (1999-02-12). "Synthesis of vinyl nonaflates derived from β-ketoesters, β-diketones or α-diketones and their palladium-catalyzed cross-coupling reactions with organozinc halides". *Tetrahedron* 55 (7): 2103–2112. doi:10.1016/S0040-4020(98)01221-6. ISSN 0040-4020.

6. Lyapkalo, Ilya M.; Matthias Webel, Hans-Ulrich Reißig (2002). "An Expedient and Stereoselective Synthesis of Alkenyl Nonaflates from Silyl Enol Ethers: Optimization, Scope and Limitations". *European Journal of Organic Chemistry* **2002** (6): 1015–1025. ISSN 1099-0690.

7. Bräse, Stefan (**1999**). "Synthesis of Bis(enolnonaflates) and their 4-exo-trig-Cyclizations by Intramolecular Heck Reactions". *Synlett* **1999** (10): 1654–1656. doi:10.1055/s-1999-2892. ISSN 0936-5214.

8. Shekhar, Shashank; Travis B. Dunn; Brian J. Kotecki; Donna K. Montavon; Steven C. Cullen (**2011**). "A General Method for Palladium-Catalyzed

Reactions of Primary Sulfonamides with Aryl Nonaflates". *J. Org. Chem.* 76 (11): 4552–4563. doi:10.1021/jo200443u. ISSN 0022-3263.

9. Uemura, Minoru; Hideki Yorimitsu; Koichiro Oshima (2008-02-18). "Cp∗Li as a base: application to palladium-catalyzed cross-coupling reaction of aryl-X or alkenyl-X (X=I, Br, OTf, ONf) with terminal acetylenes". *Tetrahedron* 64 (8): 1829–1833. doi:10.1016/j.tet. 2007.11.095. ISSN 0040-4020.

10. Quek, Ser Kiang; Ilya M. Lyapkalo; Han Vinh Huynh (**2006**-03-27). "Synthesis and properties of N,N'-dialkylimidazolium bis(nonafluoro-butane-1-sulfonyl)imides: a new subfamily of ionic liquids". *Tetrahedron* 62 (13): 3137–3145. doi:10.1016/j.tet.2006.01.015. ISSN 0040-4020.

11. Hashmi, A. Stephen K.; Tobias Hengst; Christian Lothschütz; Frank Rominger (**2010**). "New and Easily Accessible Nitrogen Acyclic Gold(I) Carbenes: Structure and Application in the Gold-Catalyzed Phenol Synthesis as well as the Hydration of Alkynes". *Advanced Synthesis & Catalysis* 352 (8): 1315–1337. doi:10.1002/adsc.201000126. ISSN 1615-4169.

3.1.32. PhenoFluor

Structure:	
IUPAC Name:	1,3-Bis(2,6-diisopropylphenyl)-2,2-difluoro-2,3-dihydro-1*H*-imidazole
Other Name:	PhenoFluor
CAS Number:	1314657-40-3
Molecular formula:	$C_{27}H_{36}F_2N_2$
Molar mass:	426.58 g/mol
Appearance:	Solid

Density:	0.856 g/mL (0.1 M solution in Toluene)
Melting point:	213 ^{0}C Decomposition
Boiling point:	-
Solubility in water:	-
Solubility:	Soluble in Toluene

Preparation and properties:

One of the simple method for the synthesis of Phenofluor.[1]

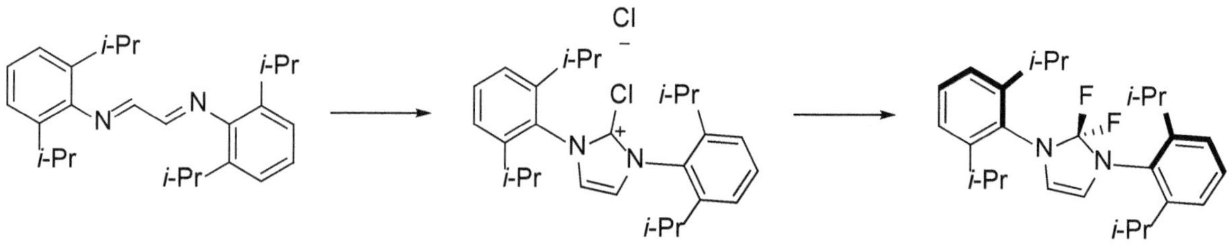

Scheme 1

Applications:

PhenoFluor reagent is utilized in deoxyfluorination of phenols, without activation of the substrate, to provide ipso substitution to the corresponding aryl fluoride.[2]

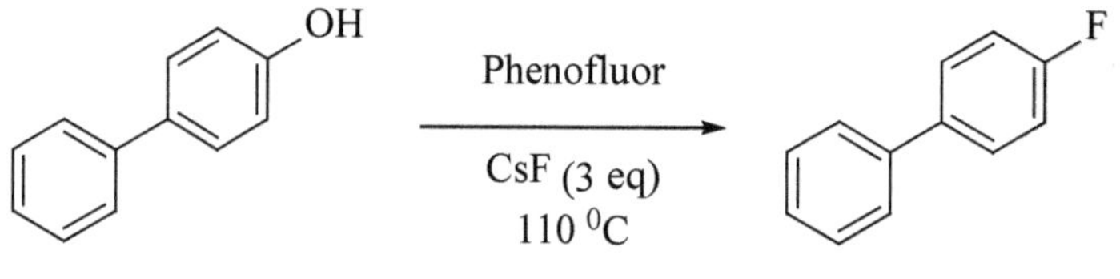

Scheme 2

Basic nitrogen-containing heteroaromatics, such as isoquinolines, quinoline, and imidazole were well tolerated in deoxyfluorination reactions (7−10). Pyrimidine 15 was obtained in 32% yield, possibly due to its high volatility.[3]

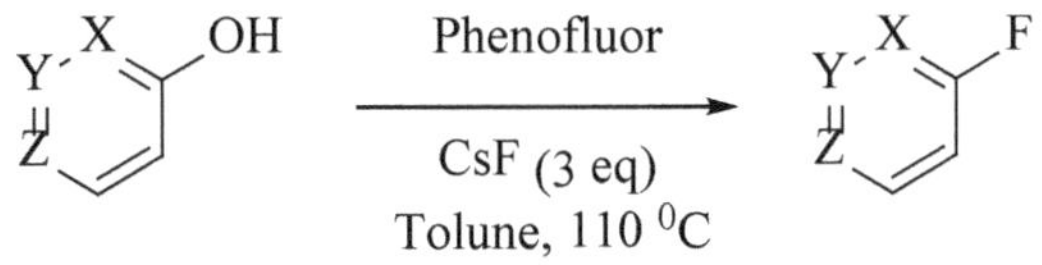

Scheme 3

PhenoFluor compared with other commercially available deoxyfluorination reagents and illustrates the fact that PhenoFluor gives access to fluorinated molecules that are practically inaccessible by deoxyfluorination using other reagents.[4]

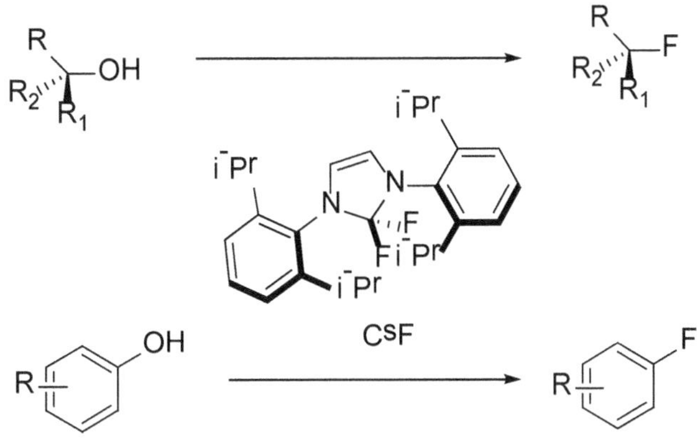

Scheme 4

PhenoFluor™ is a thermally stable deoxyfluorinating agent that delivers aryl fluorides from phenols in a one-step, high–yield, ipso substitution reaction. The method is operationally simple, regiospecific, scalable, and compatible with a variety of functional groups including amines, aldehydes, and heterocycles. More recently, it was reported that PhenoFluor™ can be used in a Late Stage Functionalization (LSF) strategy by deoxyfluorinating several natural products and drug-like molecules in a highly regioselective fashion.

Mechanism:

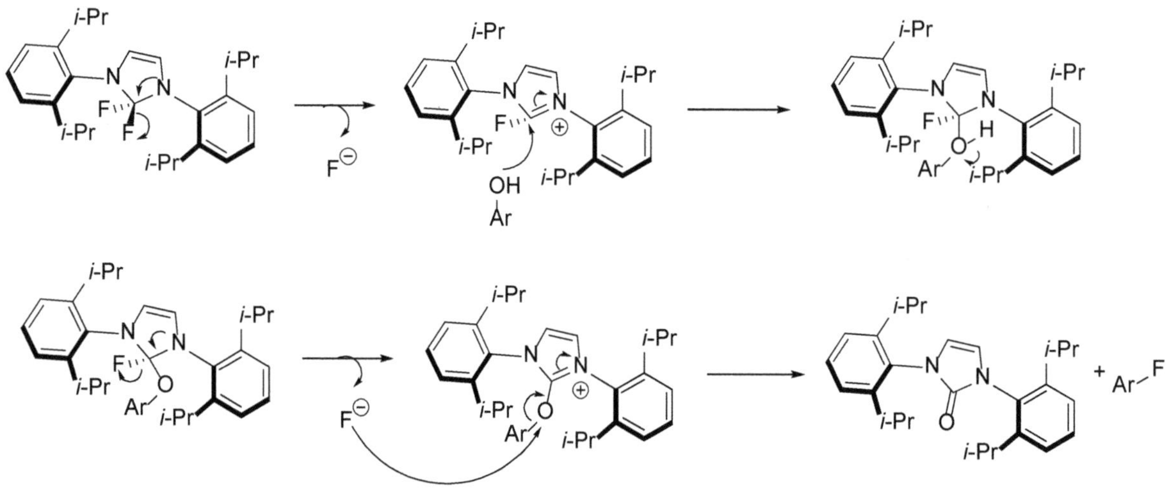

Scheme 5

Safety:

Choose body protection in relation to its type, to the concentration and amount of dangerous substances, and to the specific work-place., The type of protective equipment must be selected according to the concentration and amount of the dangerous substance at the specific workplace.

Avoid dust formation. Avoid breathing vapours, mist or gas. Do not let product enter drains. Sweep up and shovel. Keep in suitable, closed containers for disposal.

If breathed in, move person into fresh air. If not breathing, give artificial respiration. Wash off with soap and plenty of water. Flush eyes with water as a precaution. Never give anything by mouth to an unconscious person. Rinse mouth with water.

Reference:

1. Tang, P.; Wang, W.; Ritter, T. *Journal of the American Chemical Society.* 133(30), 11482-11484 (**2011**).
2. *Organic Process Research & Development.* **2014**, 18, 1041 – 1044.
3. *Journal of American Chemical Society.* **2013**, 135, 2470 – 2473.
4. Sladojevich, F.; Arlow, S. I.; Tang, P.; Wang, W.; Ritter, T. *Journal of American Chemical Society.* **2012**, ASAP. DOI: 10.1021/ja3125405.

3.1.33. Platinum hexafluoride

Platinum hexafluoride is the chemical compound with the formula PtF_6. It is a dark-red volatile solid that forms a red gas. The compound is a unique example of platinum in the +6 oxidation state. With only four d-electrons, it is paramagnetic with a tripletground state. PtF_6 is a strong oxidant and a strong fluorinating agent. PtF_6 is octahedral in both the solid state and in the gaseous state. The Pt-F bond lengths are 185 picometers.[1]

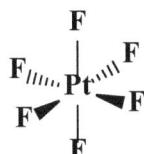

Structure:

IUPAC Name: Platinum hexafluoride

Other Name: Platinum(VI) fluoride

CAS Number: 13693-05-5

Molecular formula: PtF_6

Molar mass: 309.07 g/mol

Appearance: Dark-red crystals

Density: 3.83 g/cm^3

Melting point: 61.3 °C (142.3 °F; 334.4 K)

Boiling point: 69.14 °C (156.45 °F; 342.29 K)

Solubility in water: Reacts violently

Preparation and properties:

PtF_6 was first prepared by reaction of fluorine with platinum metal.[2] This route remains the method of choice.[1]

$Pt + 3\ F_2 \rightarrow PtF_6$

PtF_6 can also be prepared by disproportionation of PtF_5. The required PtF_5 can be obtained by fluorinating $PtCl_2$:

$$2\ PtCl_2 + 5\ F_2 \rightarrow 2\ PtF_5 + 2\ Cl_2$$

$$2\ PtF_5 \rightarrow PtF_6 + PtF_4$$

Hexafluoroplatinates

Platinum hexafluoride can gain an electron to form the hexafluoroplatinate anion, PtF_6^-. It is formed by reacting platinum hexafluoride with relatively uncationisable elements and compounds, for example with xenon to form "$XePtF_6$" (actually a mixture of $XeFPtF_5$, $XeFPt_2F_{11}$, and $Xe_2F_3PtF_6$), known as xenon hexafluoroplatinate. The discovery of this reaction in 1962 proved that noble gasesform chemical compounds. Previous to the experiment with xenon, PtF_6 had been shown to react with oxygen to form $[O_2]^+[PtF_6]^-$, dioxygenyl hexafluoroplatinate.

References:

1. Drews, T.; Supel, J.; Hagenbach, A.; Seppelt, K. "Solid State Molecular Structures of Transition Metal Hexafluorides". *Inorganic Chemistry* **2006**, vol. 45, pp 3782-3788. doi:10.1021/ic052029f.

2. Weinstock, B.; Claassen, H. H.; Malm, J. G. (**1957**). "Platinum Hexafluoride". *Journal of the American Chemical Society* **79**: 5832–5832.doi: 10.1021/ja01578a073.

3.1.34. Potassium fluoride

Potassium fluoride is the chemical compound with the formula KF. After hydrogen fluoride, KF is the primary source of the fluoride ion for applications in manufacturing and in chemistry. It is an alkali halide and occurs naturally as the rare mineral carobbiite. Aqueous solutions of KF will etch glass due to the formation of soluble fluorosilicates, although HF is more effective.

Structure:

IUPAC Name:	Potassium fluoride
CAS Number:	7789-23-3 (anhydrous),
	13455-21-1 (dihydrate)
Molecular formula:	KF
Molar mass:	58.0967 g/mol (anhydrous),
	94.1273 g/mol (dihydrate)
Appearance:	Colourless (White powder)
Odour:	no data available
Density:	2.48 g/cm^3
Melting point:	858 °C (1,576 °F; 1,131 K) (anhydrous)
	41 °C (dihydrate)
	19.3 °C (trihydrate)
Boiling point:	1,502 °C (2,736 °F; 1,775 K)

Solubility in water:

anhydrous:
92 g/100 mL (18 °C)
102 g/100 mL (25 °C)

dehydrate:
349.3 g/100 mL (18 °C)

Solubility:	soluble in HF
	insoluble in alcohol
LD50 (Lethal dose):	245 mg/kg (oral, rat)

Preparation and properties:

Potassium fluoride is prepared by dissolving potassium carbonate in excess hydrofluoric acid. Evaporation of the solution forms crystals of potassium bifluoride. The bifluoride on heating yields potassium fluoride:

$$K_2CO_3 + 4HF \rightarrow 2KHF_2 + CO_2\uparrow + H_2O$$

$$KHF_2 \rightarrow KF + HF\uparrow$$

The salt must not be prepared in glass or porcelain vessels as HF and the aqueous solution of KF corrode glass and porcelain. Heat resistant plastic or platinum containers may be used.

Applications:

In organic chemistry, KF can be used for the conversion of chlorocarbons into fluorocarbons, via the Finkelstein reaction.[2] Such reactions usually employ polar solvents such as dimethyl formamide, ethylene glycol, and dimethyl sulfoxide.[3]

Safety:

Like other sources of the fluoride ion, F⁻, KF is poisonous, although lethal doses approach gram levels for humans. It is harmful by inhalation and ingestion. It is highly corrosive, and skin contact may cause severe burns.

References:

1. Vogel, A. I.; Leicester, J.; Macey, W. A. T. "n-Hexyl Fluoride". *Org. Synth.*; *Coll.* Vol. 4, p. 525.

2. Han, Q.; Li, H-Y. "Potassium Fluoride" in Encyclopedia of Reagents for Organic Synthesis, **2001** John Wiley & Sons, New York. doi:10.1002/047084 289X.rp214.

3.1.35. Selectfluor

1-Chloromethyl-4-fluoro-1,4-diazoniabicyclo[2.2.2]octanebis(tetrafluoroborate) or Selectfluor, a trademark of Air Products and Chemicals, is a reagent in chemistry that is used as a fluorine donor. This compound is a derivative of the heterocycle DABCO . This colourless salt was first described in 1992[1] and has since been commercialized for use in organofluorine chemistry for electrophilic fluorination.

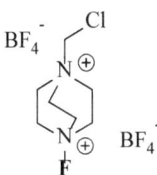

Structure:	
IUPAC Name:	1-(chloromethyl)-4-fluoro-1,4-diazo niabicyclo[2.2.2]octane ditetrafluoroborate
Other Name:	F-TEDA, N-Chloromethyl-N-fluorotriethy lenediammonium bis(tetrafluoroborate)
CAS Number:	140681-55-6
Molecular formula:	$C_7H_{14}B_2ClF_9N_2$
Molar mass:	354.26 g/mol
Appearance:	Colourless solid
Density:	-
Melting point:	234 °C
Boiling point:	decompose on heating
Solubility in water:	Reacts
LD50 (Lethal dose):	rabbit - > 2.000 mg/kg

Preparation and properties:

Selectfluor is synthesized by the N-alkylation of diazabicyclo[2.2.2]octane (DABCO) with dichloromethane, followed by ion exchange with sodium tetrafluoroborate (replacing the chloride counterion for the tetrafluoroborate). Finally, this salt is treated with elemental fluorine and sodium tetrafluoroborate:[1]

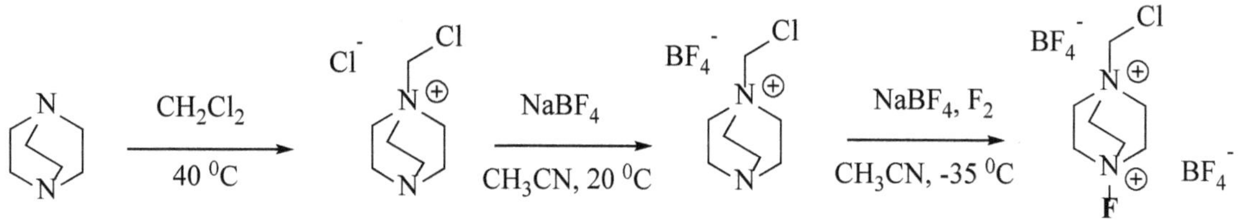

Scheme 1

Applications:

The conventional source of "electrophilic fluorine," i.e. the equivalent to the superelectrophile F^+, is gaseous fluorine, which requires specialised equipment for manipulation. Selectfluor reagent is a salt, the use of which requires only routine procedures. Like F_2, the salt delivers the equivalent of F^+. It is mainly used in the synthesis of organofluorine compounds:[2][3][4][5]

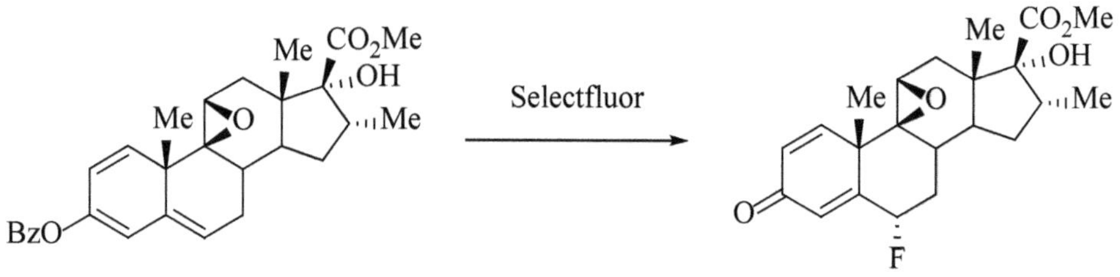

Scheme 2

Specialized applications

Selectfluor reagent also serves as a strong oxidant, a property that is useful in other reactions in organic chemistry. Oxidation of alcohols and phenols. As applied to electrophilic iodination, Selectfluor reagent activates the I_2 bond.

Nucleophilic substrates

Simple fluorinations of alkenes often produce complex mixtures of products. However, cofluorination in the presence of a nucleophile proceeds cleanly to give vicinal alkoxyfluorides.[6] Alkynes are not fluorinated with N-F reagents. A surfactant was used to facilitate contact between aqueous Selectfluor and the alkene.

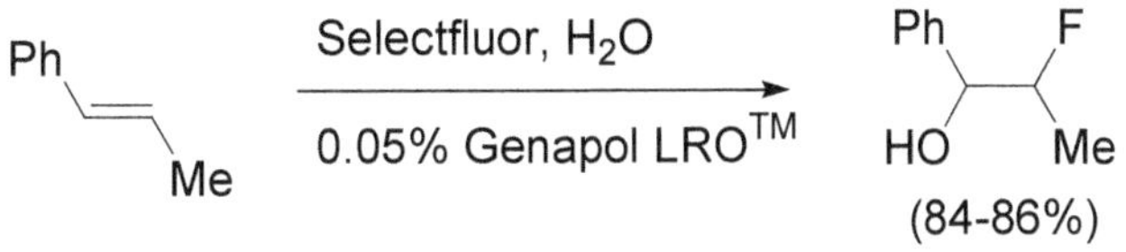

Scheme 3

Fluorination of electron-rich aromatic compounds gives aryl fluorides. The two most common problems in this class of reactions are low *ortho/para* selectivities and dearomatization (the latter is a particularly significant problem for phenols).[7]

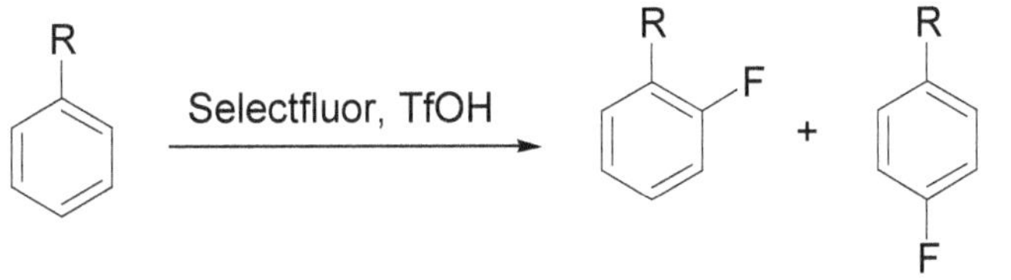

R	ortho:para
Me	68:32
OMe	45:55
F	23:77
Cl	69:31

Scheme 4

Enol ethers and glycals are nucleophilic enough to be fluorinated by Selectfluor.[8] Similar to other alkenes, cohalogenation can be accomplished either by isolation of the intermediate adduct and reaction with a nucleophile or direct displacement of DABCO *in situ*. Enols can be fluorinated enantioselectively (see above) in the presence of a chiral fluorinating agent.

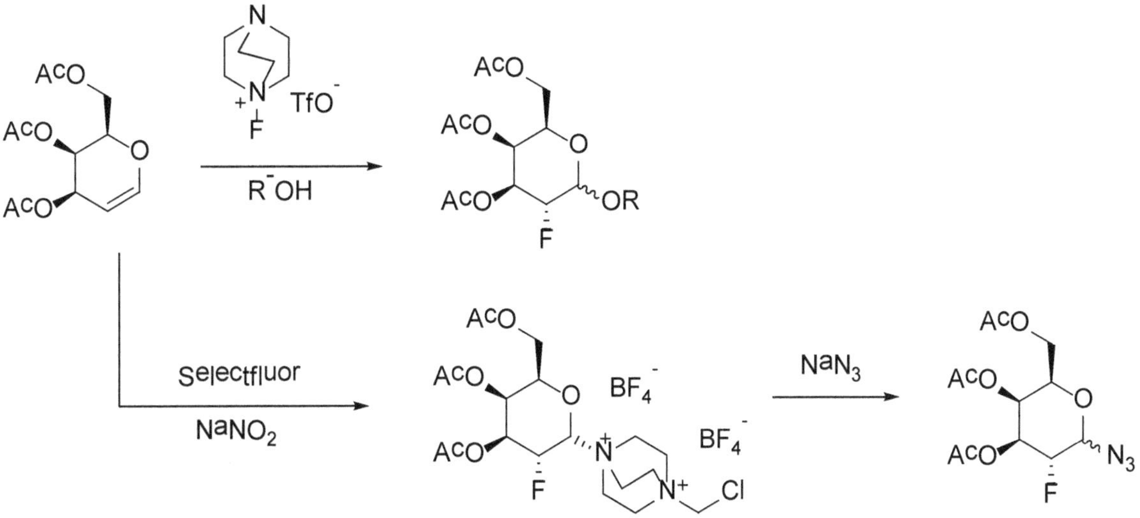

Scheme 5

Safety:

Avoid contact with skin and eyes. Avoid formation of dust and aerosols. Face shield and safety glasses Use equipment for eye protection tested and approved under appropriate government standards such as NIOSH (US) or EN 166(EU). Handle with gloves.

References:

1. Banks, R. Eric; Mohialdin-Khaffaf, Suad N.; Lal, G. Sankar; Sharif, Iqbal; Syvret, Robert G. (**1992**). "1-Alkyl-4-fluoro-1,4-diazoniabicyclo [2.2.2]octane salts: a novel family of electrophilic fluorinating agents". *Journal of the Chemical Society Chemical Communications* (8): 595. doi:10.1039/C3992 0000595.

2. Banks, R. Eric; Besheesh, Mohamed K.; Mohialdin-Khaffaf, Suad N.; Sharif, Iqbal (**1996**). "*N*-Halogeno compounds. Part 18. 1-Alkyl-4-fluoro-1,4-diazoniabicyclo[2.2.2]octane salts: user-friendly site-selective electrophilic fluorinating agents of the *N*-fluoroammonium class". *Journal of the Chemical Society Perkin 1*: 2069–2076. doi:10.1039/P19960002069.

3. Manral, Laxmi (**2006**). "Selectfluor (F-TEDA-BF$_4$) C$_7$H$_{14}$B$_2$ClF$_9$N$_2$". *Synlett* (5): 0807. doi:10.1055/s-2006-933124.

4. Stojan Stavbera and Marko Zupana (**2005**). "Selectfluortm F-TEDA-BF4 As a Versatile Mediator or Catalyst in Organic Chemistry". *Acta Chim. Slov.* **52**: 13–26.

5. Singh, R.P.; Shreeve, J. M. (**2004**). "Recent Highlights in Electrophilic Fluorination with 1-Chloromethyl-4-Fluoro-1,4-Diazoniabicyclo[2.2.2] Octane Bis(Tetrafluoroborate)". *Acc. Chem. Res.* **37** (1): 31–44. doi:10.1021/ar0300 43v. PMID 14730992.

6. Lal, G. S. *J. Org. Chem.* **1993**, 58, 2791.

7. Zupan, M.; Iskra, J.; Stavber, S. *Bull. Chem. Soc. Jpn.* **1995**, *68*, 1655.

8. Albert, M.; Dax, K.; Ortner, J. *Tetrahedron* **1998**, *54*, 4839.

3.1.36. Selenium tetrafluoride

Selenium tetrafluoride (SeF_4) is an inorganic compound. It is a colourless liquid that reacts readily with water. It can be used as a fluorinating reagent in organic syntheses (fluorination of alcohols, carboxylic acids or carbonyl compounds) and has advantages over sulfur tetrafluoride in that milder conditions can be employed and it is a liquid rather than a gas.

Structure:

IUPAC Name:	Selenium hexafluoride
Other Name:	Selenium fluoride
CAS Number:	13465-66-2
Molecular formula:	SeF_4
Molar mass:	154.954 g/mol
Appearance:	Colourless liquid
Density:	2.77 g/cm^3
Melting point:	−13.2 °C (8.2 °F; 259.9 K)
Boiling point:	101 °C (214 °F; 374 K)
Solubility in water:	-

Preparation and properties:

The first reported synthesis of selenium tetrafluoride was by Paul Lebeau in 1907, who treated selenium with fluorine:[1]

$$Se + 2\,F_2 \rightarrow SeF_4$$

A synthesis involving more easily handled reagents entails the fluorination of selenium dioxide with sulfur tetrafluoride:[2]

$$SF_4 + SeO_2 \rightarrow SeF_4 + SO_2$$

An intermediate in this reaction is seleninyl fluoride ($SeOF_2$).

Other methods of preparation include fluorinating elemental selenium with chlorine trifluoride:

$$3 \; Se + 4 \; ClF_3 \rightarrow 3 \; SeF_4 + 2 \; Cl_2$$

Reactions:

In HF, SeF_4 behaves a weak base, weaker than sulfur tetrafluoride, SF_4 ($K_b = 2 \times 10^{-2}$):

$$SeF_4 + HF \rightarrow SeF_3^+ + HF_2^-; \; (K_b = 4 \times 10^{-4})$$

Ionic adducts containing the SeF_3^+ cation are formed with SbF_5, AsF_5, NbF_5, TaF_5, and BF_3.[3] With caesium fluoride, CsF, the SeF_5^- anion is formed, which has a square pyramidal structure similar to the isoelectronic chlorine pentafluoride, ClF_5 and bromine pentafluoride, BrF_5.[4] With 1,1,3,3,5,5-hexamethylpiperidinium fluoride or 1,2-dimethylpropyltrimethylammonium fluoride, the SeF_6^{2-} anion is formed. This has a distorted octahedral shape which contrasts to the regular octahedral shape of the analogous $SeCl_6^{2-}$. [5]

Applications:

Use of the selenium tetrafluoride as a fluorinating reagents example is given below. Cyclohexanecarbaldehyde reacts with selenium tetrafluoride to gives (difluoromethyl)cyclohexane.[6]

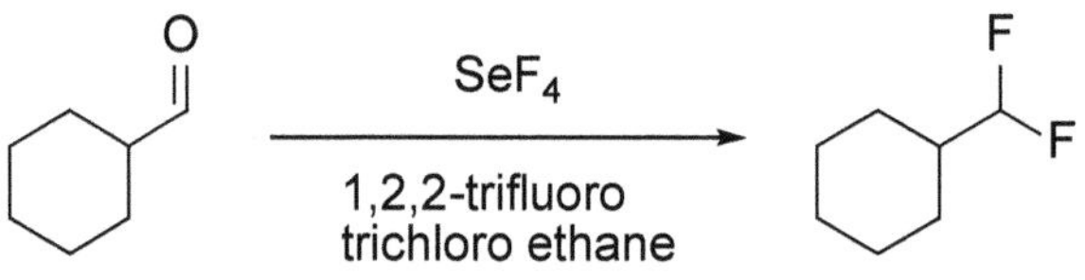

Mercury reacts with selenium tetrafluoride gives dimercury difluoride.[7]

$$Hg \; + \; SeF_4 \xrightarrow{\text{Neat}} F_2Hg_2$$

Structure and Bonding:

Selenium in SeF_4 has an oxidation state of +4. Its shape in the gaseous phase is similar to that of SF_4, having a see-saw shape.VSEPR theory predicts a pseudo-trigonal pyramidal disposition of the five electron pairs around the selenium atom. The axial Se-F bonds are 177 pm with an F-Se-F bond angle of 169.2°. The two other fluorine atoms are attached by shorter bonds (168 pm), with an F-Se-F bond angle of 100.6°. In solution at low concentrations this monomeric structure predominates, but at higher concentrations evidence suggests weak association between SeF_4 molecules leading to a distorted octahedral coordination around the selenium atom. In the solid the selenium center also has a distorted octahedral environment.

Safety:

Selenium hexafluoride is corrosive. Contact with the eyes may cause severe irritation, possibly leading to burns and permanent eye damage. Selenium hexafluoride is corrosive to the skin. Contact may cause severe irritation, leading to burns and irreversible skin damage. Selenium hexafluoride may cause skin sensitization. Selenium is toxic so avoid the contact and breath. Wash skin immediately with water for at least 15 minutes and then soak in 0.2% Hyamine solution or 13% Zephiran for 1 to 4 hours, depending upon the severity of the burns. Seek medical attention.

References:

1. Paul Lebeau (**1907**). "Action of Fluorine on Selenium Tetrafluoride of Selenium". *Comptes Rendus Acad. Sci., Paris* 144: 1042.

2. Konrad Seppelt, Dieter Lentz, Gerhard Klöter "Selenium Tetrafluoride, Selenium Difluoride Oxide (Seleninyl Fluoride), and Xenon Bis[Pentafluorooxoselenate(VI)]" *Inorg. Synth.* **1987**, vol. 24, 27-31. doi:10.1002/9780470132555.ch9.

3. R. J. Gillespie; A. Whitla (**1970**). "Selenium tetrafluoride adducts. II. Adducts with boron trifluoride and some pentafluorides". *Can. J. Chem.* 48 (4): 657–663. doi:10.1139/v70-106.

4. K. O. Christe, E. C. Curtis, C. J. Schack, D. Pilipovich (**1972**). "Vibrational Spectra and Force Constants of the Square-Pyramidal Anions SF_5^-, SeF_5^-, and TeF_5^-". *Inorganic Chemistry* 11(7): 1679–1682. doi:10.1021/ic50113a046.

5. Ali Reza Mahjoub, Xiongzhi Zhang, Konrad Seppelt (**1995**). "Reactions of the Naked Fluoride Ion: Syntheses and Structures of SeF_6^{2-} and BrF_6^-". *Chemistry - A European Journal* **1** (4): 261–265. doi:10.1002/chem.19950010410.

6. *Journal of American Chemical Society.* **1974**, Vol. 96, #3, p. 925-927.

7. *Gmeli Handbook.* Hg: M Vol. 82, 2, p. 403-404.

3.1.37. Silver(II) fluoride

Silver(II) fluoride is a chemical compound with the formula AgF_2. It is a rare example of a silver(II) compound. Silver usually exists in its +1 oxidation state. It is used as a fluorinating agent.

Structure:	F—Ag—F
IUPAC Name:	silver(II) fluoride
Other Name:	silver difluoride
CAS Number:	7783-95-1
Molecular formula:	AgF_2
Molar mass:	145.865 g/mol
Appearance:	white or grey crystalline powder, hygroscopic
Density:	4.58 g/cm^3

Melting point:	690 °C (1,274 °F; 963 K)
Boiling point:	700 °C (1,292 °F; 973 K) (decomposes)
Solubility in water:	Decomposes violently

Preparation and properties:

AgF_2 can be synthesized by fluorinating Ag_2O with elemental fluorine. Also, at 200 °C (473 K) elemental fluorine will react with AgF or AgCl to produce AgF_2.[1][2]

As a strong fluorinating agent, AgF_2 should be stored in Teflon or a passivated metal container. It is light sensitive.

AgF_2 can be purchased from various suppliers, the demand being less than 100 kg/year. While laboratory experiments find use for AgF_2, it is too expensive for large scale industry use. In 1993, AgF_2 cost between 1000-1400 US dollars per kg.

Composition and structure:

AgF_2 is a white crystalline powder, but it is usually black/brown due to impurities. The F/Ag ratio for most samples is < 2, typically approaching 1.75 due to contamination with Ag and oxides and carbon.[3]

For some time, it was doubted silver was actually in the +2 oxidation state rather in some combination of states such as $Ag^I[Ag^{III}F_4]$, which would be similar to silver(I,III) oxide. Neutron diffraction studies, however, confirmed its description as silver(II). The $Ag^I[Ag^{III}F_4]$ was found to be present at high temperatures, but it was unstable with respect to AgF_2.[4]

In the gas phase, AgF_2 is believed to have $D_{\infty h}$ symmetry.

Approximately 14 kcal/mol (59 kJ/mol) separate the ground and first states. The compound is paramagnetic, but it becomes ferromagnetic at temperatures below −110 °C (163 K).

Applications:

AgF_2 is a strong fluorinating and oxidising agent. It is formed as an intermediate in the catalysis of gaseous reactions with fluorine by silver. With fluoride ions, it forms complex ions such as AgF−3, the blue-violet AgF_2^{-4}, and AgF_4^{-6}.[5]

It is used in the fluorination and preparation of organic perfluorocompounds.[6] This type of reaction can occur in three different ways (here Z refers to any element or group attached to carbon, X is a halogen):

$CZ_3H + 2\ AgF_2 \rightarrow CZ_3F + HF + 2\ AgF$

$CZ_3X + 2AgF_2 \rightarrow CZ_3F + X_2 + 2\ AgF$

$Z_2C=CZ_2 + 2\ AgF_2 \rightarrow Z_2CFCFZ_2 + 2\ AgF$

Similar transformations can also be effected using other high valence metallic fluorides such as CoF_3, MnF_3, CeF_4, and PbF_4.

AgF_2 is also used in the fluorination of aromatic compounds, although selective monofluorinations are more difficult:[7]

$C_6H_6 + 2\ AgF_2 \rightarrow C_6H_5F + 2\ AgF + HF$

AgF_2 oxidises xenon to xenon difluoride in anhydrous HF solutions.[8]

$2\ AgF_2 + Xe \rightarrow 2\ AgF + XeF_2$

It also oxidises carbon monoxide to carbonyl fluoride.

$2\ AgF_2 + CO \rightarrow 2\ AgF + COF_2$

It reacts with water to form oxygen gas:[citation needed]

$4\ AgF_2 + 4\ H_2O \rightarrow 2\ Ag_2O + 8\ HF + O_2$

Recently Hartwig showed that AgF_2 can be used to selecively fluorinate pyridine at the ortho position under mild conditions.[9]

Safety:

AgF_2 is a very strong oxidizer that reacts violently with water,[10] reacts with dilute acids to produce ozone, oxidizes iodide to iodine,[10][11] and upon contact with acetylene forms the contact explosive silver acetylide.[12] It is light-sensitive,[10] very hygroscopic and corrosive. It decomposes violently on contact with hydrogen peroxide, releasing oxygen gas.[12] It also liberates HF, F_2, and elemental silver.[11]

References:

1. Priest, H. F.; Swinehert, Carl F. (**1950**). "Anhydrous Metal Fluorides". *Inorg. Synth 3.* P.171–183. doi:10.1002/978070132340. ISBN 978-0-470-13234-0.

2. *Encyclopedia of Chemical Technology.* Kirk-Othermer. Vol.11, 4th Ed. (**1991**).

3. J.T. Wolan, G.B. Hoflund (**1998**). "Surface Characterization Study of AgF and AgF_2 Powders Using XPS and ISS". *Applied Surface Science* 125 (3–4): 251. doi:10.1016/S0169-4332(97)00498-4.

4. Hans-Christian Miller, Axel Schultz, and Magdolna Hargittai (**2005**). "Structure and Bonding in Silver Halides. A Quantum Chemical Study of the Monomers: Ag_2X, AgX, AgX_2, and AgX_3(X = F, Cl, Br, I)". *J. Am. Chem. Soc.* 127 (22): 8133–45. doi:10.1021/ja051442j. PMID 15926841.

5. Egon Wiberg; Nils Wiberg; Arnold Frederick Holleman (**2001**). *Inorganic chemistry.* Academic Press. pp. 1272–1273. ISBN 0-12-352651-5.

6. Rausch, D.; Davis, r.; Osborne, D. W. (**1963**). "The Addition of Fluorine to Halogenated Olefins by Means of Metal Fluorides". *J. Org. Chem.* 28 (2): 494–497. doi:10.1021/jo01037a055.

7. Zweig, A.; Fischer, R. G.; Lancaster, J. (**1980**). "New Methods for Selective Monofluorination of Aromatics Using Silver Difluoride". *J. Org. Chem.* **45** (18): 3597. doi:10.1021/jo01306a011.

8. Levec, J.; Slivnik, J.; Zemva, B. (**1974**). "On the Reaction Between Xenon and Fluorine". *Journal of Inorganic Nuclear Chemistry* 36 (5): 997. doi:10.1016/ 0022-1902(74)80203-4.

9. Fier, P. S.; Hartwig, J. F. (**2013**). "Selective C-H Fluorination of Pyridines and Diazines Inspired by a Classic Amination Reaction". *SCIENCE* 342: 956. doi:10.1126/science.1243759.

10. Dale L. Perry; Sidney L. Phillips (**1995**). *Handbook of inorganic compounds*. CRC Press. p. 352. ISBN 0-8493-8671-3.

11. W. L. F. Armarego; Christina Li Lin Chai (**2009**). *Purification of Laboratory Chemicals* (6th ed.). Butterworth-Heinemann. p. 490. ISBN 1-85617-567-7.

12. Richard P. Pohanish; Stanley A. Greene (**2009**). *Wiley Guide to Chemical Incompatibilities* (3rd ed.). John Wiley and Sons. p. 93. ISBN 0-470-38763-7.

3.1.38. Sulfur tetrafluoride

Sulfur tetrafluoride is the chemical compound with the formula SF_4. This species exists as a gas at standard conditions. It is a corrosive species that releases dangerous HF upon exposure to water or moisture. Despite these unwelcome characteristics, this compound is a useful reagent for the preparation of organo-fluorine compounds,[1] some of which are important in the pharmaceutical and specialty chemical industries.

Structure:

IUPAC Name: Sulfur(IV) fluoride

Other Name: Sulfur tetrafluoride

CAS Number: 7783-60-0

Molecular formula: SF_4

Molar mass: 108.07 g/mol

Appearance: Colourless gas

Density: 1.95 g/cm^3, −78 °C

Melting point:	−121.0 °C
Boiling point:	−38 °C
Solubility in water:	Reacts

Preparation and properties:

SF_4 is produced by the reaction of SCl_2, Cl_2, and NaF:

$$SCl_2 + Cl_2 + 4\,NaF \rightarrow SF_4 + 4\,NaCl$$

Treatment of SCl_2 with NaF also affords SF_4, not SF_2. SF_2 is unstable, it condenses with itself to form SF_4 and SSF_2.[2]

Applications:

In organic synthesis, SF_4 is used to convert COH and C=O groups into CF and CF_2 groups, respectively.[3] Certain alcohols readily give the corresponding fluorocarbon. Ketones and aldehydes give geminal difluorides. The presence of protons alpha to the carbonyl leads to side reactions and diminished (30–40%) yield. Also diols can give cyclic sulfite esters, $(RO)_2SO$. Carboxylic acids convert to trifluoromethyl derivatives. For example treatment of heptanoic acid with SF_4 at 100-130 °C produces 1,1,1-trifluoroheptane. The coproducts from these fluorinations, including unreacted SF_4 together with SOF_2 and SO_2, are toxic but can be neutralized by their treatment with aqueous KOH.

The use of SF_4 is being superseded in recent years by the more conveniently handled diethylaminosulfur trifluoride, Et_2NSF_3, DAST, where Et= CH_3CH_2.[4] This reagent is prepared from SF_4:[5]

$$SF_4 + Me_3SiNEt_2 \rightarrow Et_2NSF_3 + Me_3SiF$$

Sulfur tetrafluoride widely used in the conversion of carboxylic acid to acid trifluoride.[6][7]

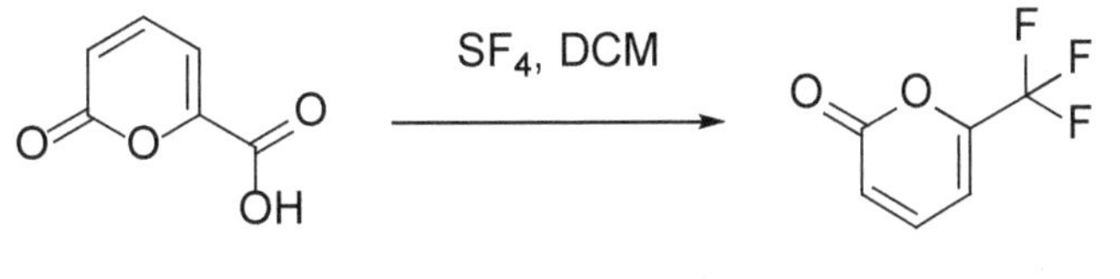

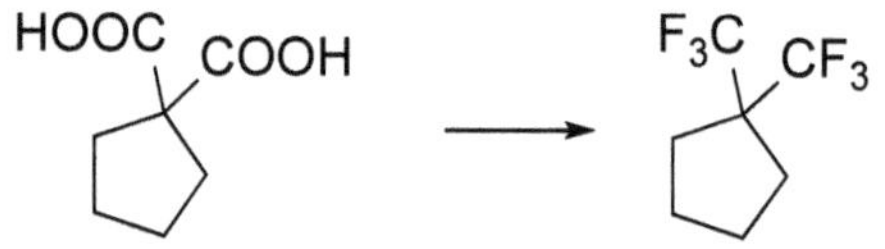

Scheme 1

Sulfur tetrafluoride is used for the conversion of alcohol to fluoride, aldehyde to difluoride and carboxylic acid to trifluoride.[8]

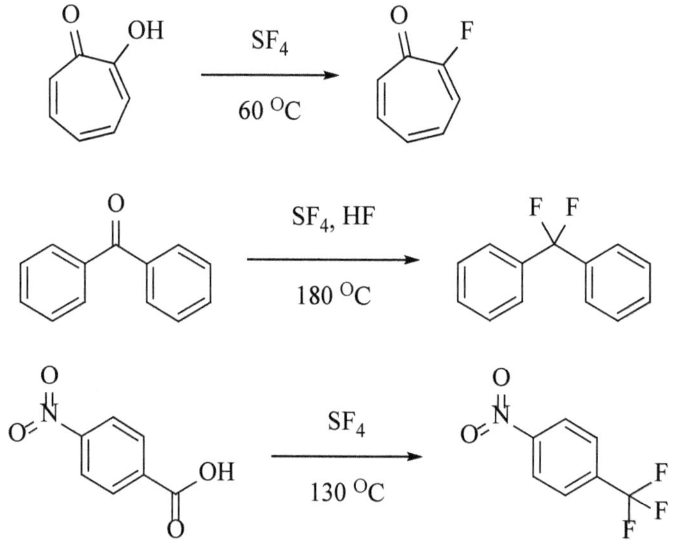

Scheme 2

Mechanism:

As per reaction mechanism during reaction HF is generated so need to handle carefully.

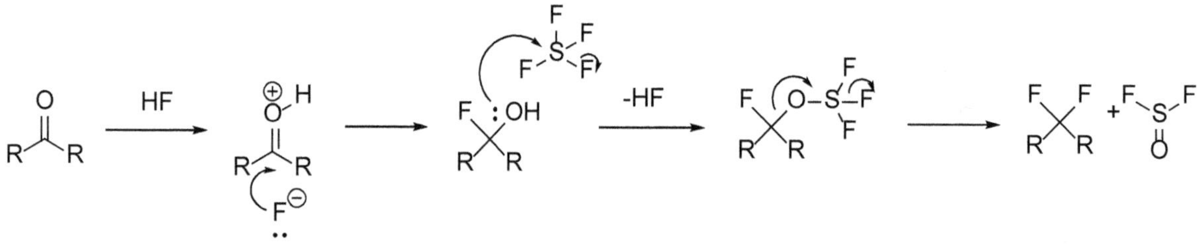

Scheme 3

Reaction:

Sulfur chloride pentafluoride (SF_5Cl), a useful source of the SF_5 group, is prepared from SF_4.[9]

Hydrolysis of SF_4 gives sulfur dioxide:[10]

$$SF_4 + 2\ H_2O \rightarrow SO_2 + 4\ HF$$

This reaction proceeds via the intermediacy of thionyl fluoride, which usually does not interfere with the use of SF_4 as a reagent.[2]

Structure:

Sulfur in SF_4 is in the formal +4 oxidation state. Of sulfur's total of six valence electrons, two form a lone pair. The structure of SF_4 can therefore be anticipated using the principles of VSEPR theory: it is a see-saw shape, with S at the center. One of the three equatorial positions is occupied by a nonbonding lone pair of electrons. Consequently, the molecule has two distinct types of F ligands, two axial and two equatorial. The relevant bond distances are $S-F_{ax} = 164.3$ pm and $S-F_{eq} = 154.2$ pm. It is typical for the axial ligands in hypervalent molecules to be bonded less strongly. In contrast to SF_4, the related molecule SF_6 has sulfur in the 6+ state, no valence electrons remain nonbonding on sulfur, hence the molecule adopts a highly symmetrical octahedral structure. Further contrasting with SF_4, SF_6 is extraordinarily inert chemically.

The ^{19}F NMR spectrum of SF_4 reveals only one signal, which indicates that the axial and equatorial F atom positions rapidly interconvert via pseudorotation.[11]

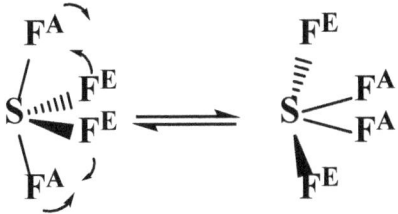

Intramolecular dynamic equilibration of SF_4.

Safety:

Disulfur decafluoride is a colourless gas or liquid with a sulfur-dioxide-like odour.[12] It is about 4 times as poisonous as phosgene. Its toxicity is thought to be caused by its disproportionation in the lungs, according to the following reaction:

$$S_2F_{10} \rightarrow SF_6 + SF_4$$

where SF_6 is inert, and SF_4 reacts inside the lungs with moisture:[13]

$$SF_4 + 2\ H_2O \rightarrow SO_2 + 4\ HF$$

References:

1. C. L. J. Wang, "Sulfur Tetrafluoride" in Encyclopedia of Reagents for Organic Synthesis (Ed: L. Paquette) **2004**, J. Wiley & Sons, New York. doi:10.1002/ 047084289.

2. F. S. Fawcett, C. W. Tullock, "Sulfur (IV) Fluoride: (Sulfur Tetrafluoride)" *Inorganic Syntheses,* **1963**, vol. 7, pp 119–124. doi:10.1002/9780470132388.

3. Hasek, W. R. "1,1,1-Trifluoroheptane". *Org. Synth.*; *Coll. Vol.* 5, p. 1082.

4. A. H. Fauq, "N,N-Diethylaminosulfur Trifluoride" in Encyclopedia of Reagents for Organic Synthesis (Ed: L. Paquette) **2004**, J. Wiley & Sons, New York. doi:10.1002/047084289.

5. W. J. Middleton, E. M. Bingham. "Diethylaminosulfur Trifluoride". *Org. Synth.*; *Coll. Vol.* 6, p. 440.

6. *Journal of Medicinal Chemistry.* **1986**, Vol. 29, # 7, p.1159-1163.

7. *Journal of Fluorine chemistry.* Vol. 102, # 1-2, p.141-146.

8. Hasek, W. R.; Smith, W. C.; Engelhardt, V. A. *J. Am. Chem. Soc.* **1960**, Vol. 82, 543.

9. Nyman, F., Roberts, H. L., Seaton, T. *Inorganic Syntheses.* **1966**, Vol. 8, p. 160 McGraw-Hill Book Company, Inc., 1966, doi:10.1002/ 9780470132395.

10. Greenwood, Norman N.; Earnshaw, Alan (**1997**). *Chemistry of the Elements* (2nd ed.). Butterworth-Heinemann. ISBN 0080379419.

11. Holleman, A. F.; Wiberg, E. "Inorganic Chemistry" Academic Press: San Diego, 2001. ISBN 0-12-352651-5.

12. "Sulfur Pentaflu". *1988 OSHA PEL Project*. CDC NIOSH.

13. Johnston, H. (**2003**). *A Bridge not Attacked: Chemical Warfare Civilian Research During World War II*. World Scientific. pp. 33–36. ISBN 981-238-153-8.

3.1.39. Togni reagents

Togni's reagents are effective for the trifluoromethylation , the design and synthesis of electrophilic trifluoromethylating reagents have been extensively researched in both academia and industry, due to the significant unique features that trifluoromethylated compounds have in pharmaceuticals, agricultural chemicals, and functional materials. In 2006 Togni and co-workers reported a new family of hypervalent iodine compounds in which the CF_3 group is bonded directly to the iodine atom.

Structure:	
IUPAC Name:	3,3-dimethyl-1-(trifluoromethyl)-1,3-dihydro-1λ^3-benzo[d][1,2]iodaoxole & 1-(trifluoromethyl)-1λ^3-benzo[d][1,2]iodaoxol-3(1H)-one
Other Name:	Togni's reagents, Trifluoromethyl-1,2-benziodoxol-3-(1*H*)-one
CAS Number:	887144-97-0
Molecular formula:	$C_{10}H_{10}F_3IO$ & $C_8H_4F_3IO_2$
Molar mass:	329.97 & 315.92 g/mol
Appearance:	Solid

Density: -

Melting point: 75-79 °C & 150-158 °C

Solubility in water: reacts

Solubility: soluble in Dichloromethae

Preparation and properties:

The overall synthetic protocol depends on a formal umpolung of the CF_3 group since nucleophilic ligand displacement with CF_3^- at the hypervalent iodine atom is carried out during the synthesis of these CF_3^+ donor reagents. For example, reaction of the methyl ester of 2-iodosylbenzoic acid **2**, with Me_3SiCF_3 in the presence of a catalytic amount of fluoride ions in CH_3CN at ambient temperature gave 1-trifluoromethyl-1,2-benziodoxol-3-(1H)-one (**3**) in 55% yield (Scheme 1).[1] Reagents **5–7** were preferentially obtained in an improved, practical, one-pot procedure by substitution of chloro substituent in **4** by an acetoxy group followed by fluoride catalyzed substitution with Ruppert's reagent (Scheme 1).[2] These reagents are shelf-stable, non-explosive under ambient conditions but should not be heated as solid materials.

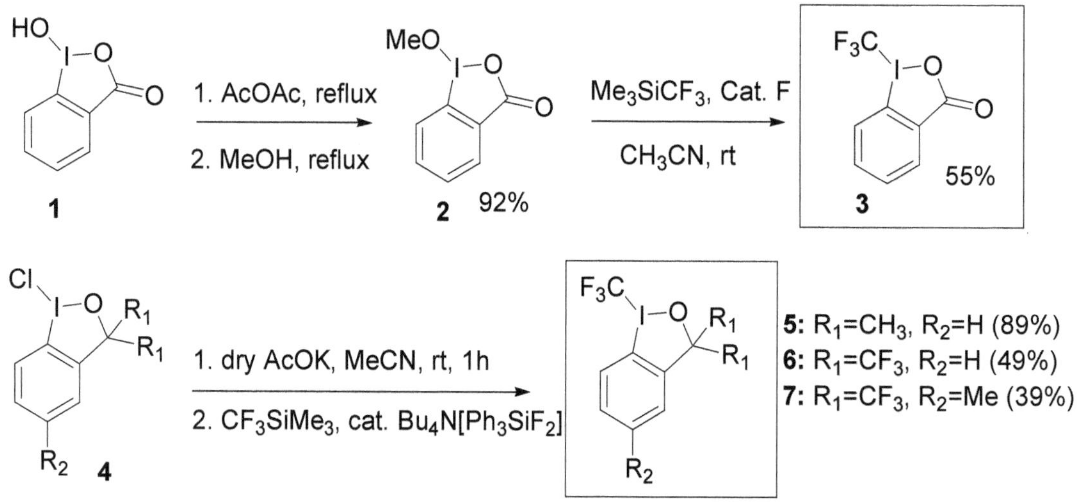

Scheme 1

Applications:

These new electrophilic trifluoromethylating reagents were initially evaluated in reactions with carbonyl compounds such as β-keto esters and α-nitro esters. In particular, reagent **5** was found to be an effective trifluoromethylating agent. Under phase-transfer catalysis the β-keto esters derived from indanone, tetralone and pentanone in the presence of **5** gave the corresponding trifluoromethylated product in 42–67% yields. The new reagents showed a clear advantage in the reaction with α-nitro esters; the reaction proceeded smoothly in CH_2Cl_2 in the presence of a catalytic amount of $CuCl_2$ (Scheme 2).[2] Interestingly, 2-(2-iodophenyl)propan-2-ol formed as by-product in the reactions of **5** with the substrates could be isolated and recycled.

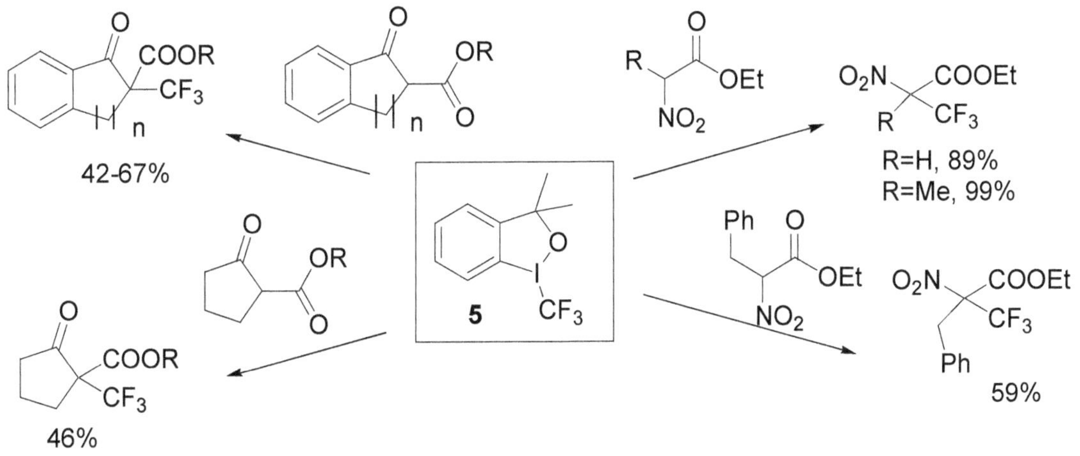

Scheme 2

The same group studied the reactivity of hypervalent iodine–CF_3 reagents with different types of sulfur-, phosphorus- and oxygen-centered nucleophiles. Firstly, it was demonstrated that sulfur-centered nucleophiles react with hypervalent iodine–CF_3 reagents. Thus, both aromatic and aliphatic thiols underwent *S*-trifluoromethylation smoothly in the presence of 1.1 equiv of **37** to afford the corresponding products in 51–99% yields (Scheme 3).[2] The reaction outperforms other methods for synthesis of the SCF_3 motif and shows high functional-group tolerance, and has particular application for the synthesis of sugar and amino acid derivatives.

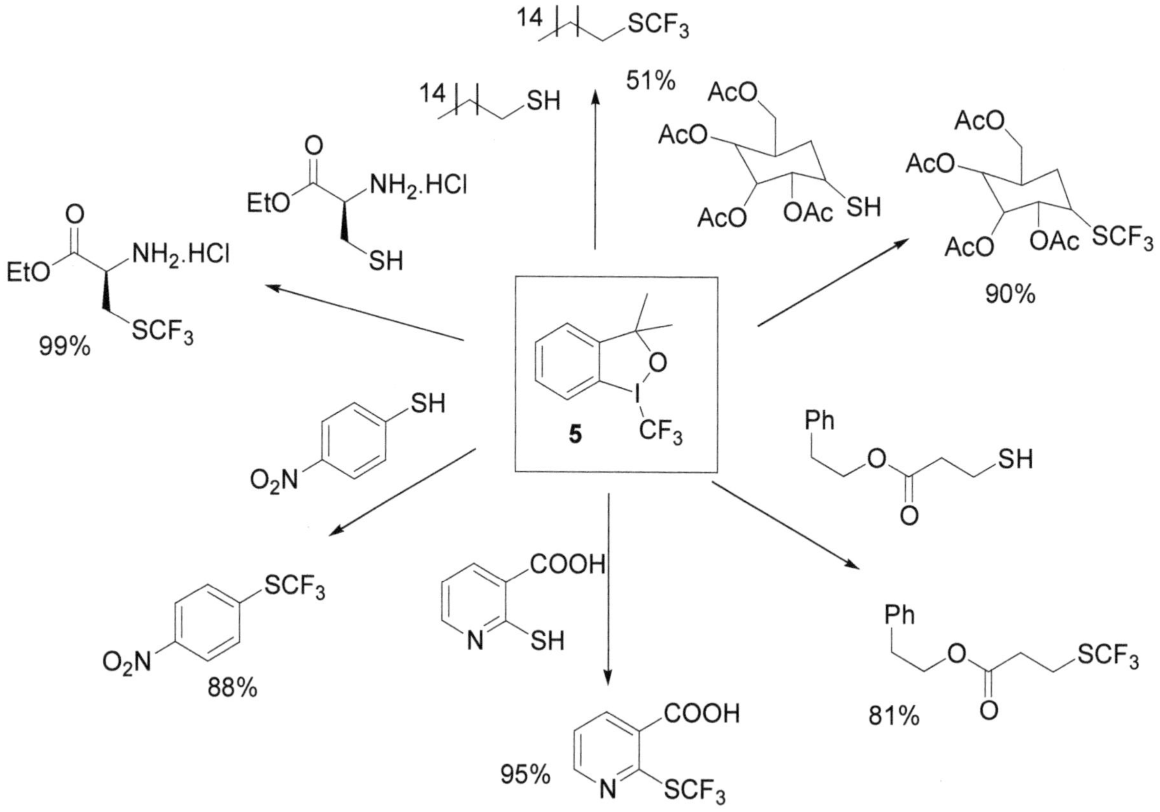

Scheme 3

3,3-Dimethyl-1-(trifluoromethyl)-1,2-benziodoxole (Togni Reagent) The direct transfer of a trifluoromethyl group usually requires harsh conditions that are often incompatible with more sensitive functionalities in a molecule. Nucleophilic trifluoromethylation is the most common method, due in large part to the broad applicability of the Ruppert–Prakash reagent (Me_3SiCF_3). Most reagents for electrophilic C- and S-trifluoromethylation are considerably less developed. Antonio Togni and coworkers have recently reported a new electrophilic reagent based on hypervalent iodine, 3,3-dimethyl- 1-(trifluoromethyl)-1,2-benziodoxole,1 which nicely complements the nucleophilic Ruppert–Prakash reagent. The Togni reagent is easy to handle, and can be exposed to moist air for short periods of time without any apparent alteration. β-Ketoesters were found to react with the Togni reagent under phase-transfer catalysis conditions to yield the α-trifluoromethylated derivatives (Scheme 4).

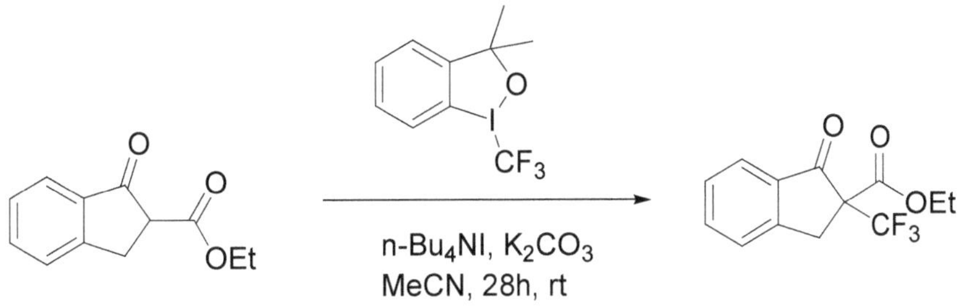

Scheme 4

More interesting are the trifluoromethylations of α-nitroesters, which yield precursors to α-trifluoromethyl-α-amino acids (Scheme 5).

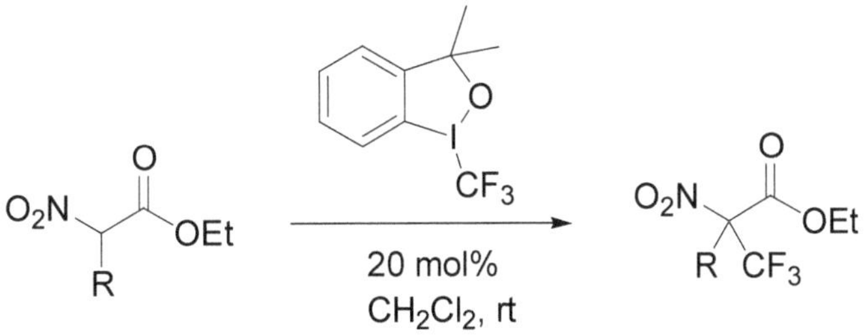

Scheme 5

Aromatic and aliphatic thiols undergo selective S-trifluoromethylation in the presence of the Togni reagent, without formation of the corresponding disulfide (Scheme 6). The reaction is remarkably tolerant of various functional groups and does not show significant solvent dependence, allowing for the use of the Togni reagent at the latter stages of syntheses of complex molecules.

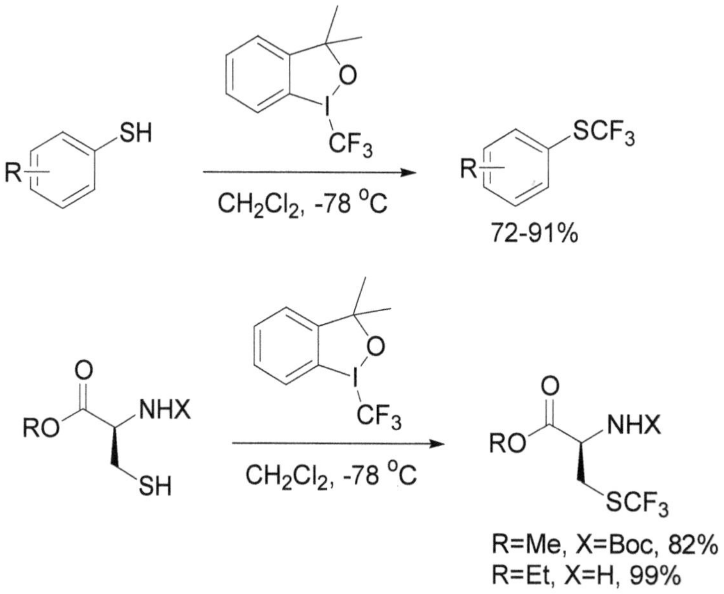

Scheme 6

Safety:

We have now found that this compound has explosive properties and would like to make the chemical community aware of this. The first hypervalent iodo intermediate in the synthesis of this reagent (1-hydroxy benz-1,2-iodoxol-(3H)3-one, 4) is also hazardous as well as probably the first Togni reagent (1,3-dihydro-3,3-dimethyl-1-(trifluoromethyl)-1,2-benziodoxole, 2).

Thus, these compounds should only be handled with the appropriate knowledge and safety measures. Specifically, laboratory work should be done behind safety shields with small amounts, open flames and circumstances that can produce sparks have to be avoided, grinding should not be done with brute force, and soft and polished tools should be used for manipulations. During preparation and especially isolation of these compounds caking should be avoided, and lumps should be dispersed early. It must be emphasized that impurities may influence the thermal and mechanical sensitiveness. The transport of explosive compounds requires the permission of competent authorities.

References:

1. Eisenberger, P.; Gischig, S.; Togni, A. *Chem.–Eur. J.* **2006**, 12, 2579–2586. doi:10.1002/chem.200501052 Return to citation in text: [1]
2. Kieltsch, I.; Eisenberger, P.; Togni, A. *Angew. Chem., Int. Ed.* **2007**, 46, 754–757. doi:10.1002/anie.200603497 Return to citation in text: [1] [2] [3]

3.1.40. Xenon difluoride

Xenon difluoride is a powerful fluorinating agent with the chemical formula XeF_2, and one of the most stable xenon compounds. Like most covalent inorganic fluorides it is moisture-sensitive. It decomposes on contact with light or (slowly) water vapour. Xenon difluoride is a dense, white crystalline solid. It has a nauseating odour and low vapour pressure.[1]

Structure:	F—Xe—F
IUPAC Name:	Xenon(II) fluoride
Other Name:	Xenone difluoride
CAS Number:	13709-36-9
Molecular formula:	F_2Xe
Molar mass:	169.29 g/mol
Appearance:	White solid
Odour:	Ozone like odour
Density:	4.32 g/cm^3, solid
Melting point:	128.6 °C (263.5 °F; 401.8 K)[2]
Boiling point:	-
Solubility in water:	25 g/l (0 °C)
Solubility:	-
Vapour pressure:	6.0×10^2 Pa

Preparation and properties:

Synthesis proceeds by the simple reaction:

$$Xe + F_2 \rightarrow XeF_2$$

The reaction needs heat, irradiation, or an electrical discharge. The product is a solid. It is purified by fractional distillation or selective condensation using a vacuum line.[2]

The first published report of XeF_2 was in October 1962 by Chernick, et al.[3] However, though published later,[4] XeF_2 was probably first created by Rudolf Hoppe at the University of Münster, Germany, in early 1962, by reacting fluorine and xenon gas mixtures in an electrical discharge.[5] Shortly after these reports, Weeks, Cherwick, and Matheson of Argonne National Laboratory reported the synthesis of XeF_2 using an all-nickel system with transparent alumina windows, in which equal parts Xe and F_2 gases react at low pressure upon irradiation by an ultraviolet source to give XeF_2.[6] Williamson reported that the reaction works equally well at atmospheric pressure in a dry Pyrex glass bulb using sunlight as a source. It was noted that the synthesis worked even on cloudy days.[7]

In the previous syntheses the F_2 reactant had been purified to remove HF. Šmalc and Lutar found that if this step is skipped the reaction rate proceeds at four times the original rate.[8]

In 1965, it was also synthesized by reacting xenon gas with dioxygen di-fluoride.[9]

Other xenon compounds may be derived from xenon difluoride. The unstable organoxenon compound $Xe(CF_3)_2$ can be made by irradiating hexafluoro-ethane to generate $CF_3\cdot$ radicals and passing the gas over XeF_2. The resulting waxy white solid decomposes completely within 4 hours at room temperature.[10]

The XeF^+ cation is formed by combining xenon difluoride with a strong fluoride acceptor, such as an excess of liquid antimony pentafluoride (SbF_5):

$$XeF_2 + SbF_5 \rightarrow XeF^+ + SbF_6^-$$

Adding xenon gas to this pale yellow solution at a pressure of 2-3 atm produces a green solution containing the paramagnetic Xe^{+2} ion,[11] which contains a Xe–Xe bond: ("apf" denotes solution in liquid SbF_5)

$$3\ Xe\ (g) + XeF+\ (apf) + SbF_5\ (l) \rightleftharpoons 2\ Xe+2\ (apf) + SbF_6^-\ (apf)$$

This reaction is reversible; removing xenon gas from the solution causes the Xe^{+2} ion to revert to xenon gas and XeF^+, and the colour of the solution returns to a pale yellow.[12]

In the presence of liquid HF, dark green crystals can be precipitated from the green solution at −30 °C:

$$Xe^{+2}\ (apf) + 4\ SbF{-}6\ (apf) \rightarrow Xe^{+2}Sb_4F_{21}^-\ (s) + 3\ F^-\ (apf)$$

X-ray crystallography indicates that the Xe-Xe bond length in this compound is 309 pm, indicating a very weak bond.[10] The Xe+2 ion is isoelectronic with the I−2 ion, which is also dark green.[13][14]

Coordination chemistry

XeF_2 can act as a ligand in coordination complexes of metals.[15] For example, in HF solution:

$$Mg(AsF_6)_2 + 4\ XeF_2 \rightarrow [Mg(XeF_2)_4](AsF_6)_2$$

Crystallographic analysis shows that the magnesium atom is coordinated to 6 fluorine atoms. Four of the fluorines are attributed to the four xenon difluoride ligands while the other two are a pair of *cis*-AsF_6^- ligands.[16]

A similar reaction is:

$$Mg(AsF_6)_2 + 2\ XeF_2 \rightarrow [Mg(XeF_2)_2](AsF_6)_2$$

In the crystal structure of this product the magnesium atom is octahedrally-coordinated and the XeF_2 ligands are axial while the AsF_6^- ligands are equatorial.

Many such reactions with products of the form $[M^x(XeF_2)_n](AF_6)_x$ have been observed, where M can be Ca, Sr, Ba, Pb, Ag, La or Nd and A can be As, Sb or P.

Recently, a compound was synthesised where a metal atom was coordinated solely by XeF_2 fluorine atoms:[17]

$$2\ Ca(AsF_6)_2 + 9\ XeF_2 \rightarrow Ca_2(XeF_2)_9(AsF_6)_4.$$

This reaction requires a large excess of xenon difluoride. The structure of the salt is such that half of the Ca^{2+} ions are coordinated by fluorine atoms from xenon difluoride, while the other Ca^{2+} ions are coordinated by both XeF_2 and AsF_6^-.

Structure:

Xenon difluoride is a linear molecule with an Xe–F bond length of 197.73 ± 0.15 pm in the vapour stage, and 200 pm in the solid phase. The packing arrangement in solid XeF_2 shows that the fluorine atoms of neighbouring molecules avoid the equatorial region of each XeF_2 molecule. This agrees with the prediction of VSEPR theory, which predicts that there are 3 pairs of non-bonding electrons around the equatorial region of the xenon atom.[15]

At high pressures, novel, non-molecular forms of xenon difluoride can be obtained. Under a pressure of ~50 GPa, XeF_2 transforms into a semiconductor consisting of XeF_4 units linked in a two-dimensional structure, like graphite. At even higher pressures, above 70 GPa, it becomes metallic, forming a three-dimensional structure containing XeF_8 units.[18] However, a recent theoretical study has put these experimental results in doubt.[19]

Applications:

1) As a fluorinating agent

Xenon difluoride is a strong fluorinating and oxidising agent.[23][24] With fluoride ion acceptors, it forms XeF^+ and $Xe2F^{+3}$ species which are even more powerful fluorinators.[15]

Among the fluorination reactions that xenon difluoride undergoes are:

- Oxidative fluorination:

$$Ph_3TeF + XeF_2 \rightarrow Ph_3TeF_3 + Xe$$

- Reductive fluorination:

$$2\ CrO_2F_2 + XeF_2 \rightarrow 2\ CrOF_3 + Xe + O_2$$

- Aromatic fluorination:

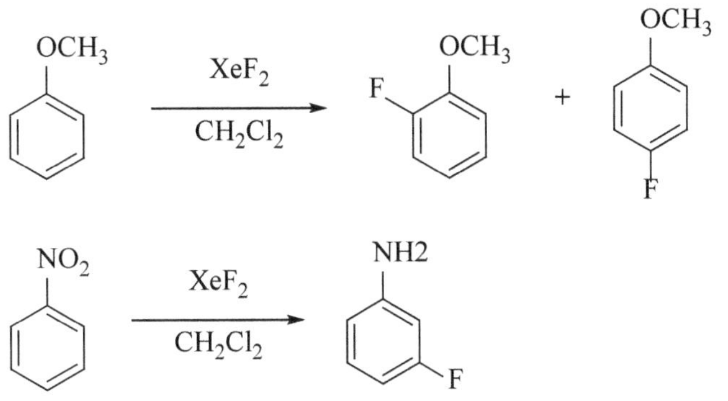

- Alkene fluorination:

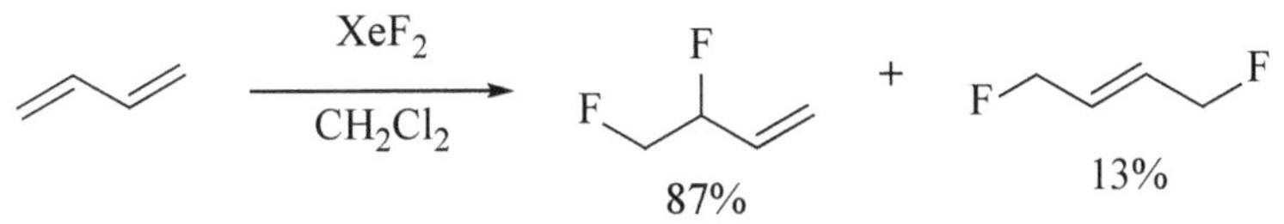

XeF_2 is selective about which atom it fluorinates, making it a useful reagent for fluorinating heteroatoms without touching other substituent's in organic compounds. For example, it fluorinates the arsenic atom in trimethylarsine, but leaves the methyl groups untouched:[20]

$$(CH_3)_3As + XeF_2 \rightarrow (CH_3)_3AsF_2 + Xe$$

XeF_2 will also oxidatively decarboxylate carboxylic acids to the corresponding fluoroalkanes:[21][22]

$$RCOOH + XeF2 \rightarrow RF + CO2 + Xe + HF$$

Silicon tetrafluoride has been found to act as a catalyst in fluorination by XeF_2.[23]

2) As an etchant

Xenon difluoride is also used as an isotropic gaseous etchant for silicon, particularly in the production of microelectromechanical systems, (MEMS), as first demonstrated in 1995.[24] Commercial systems use pulse etching with an expansion chamber [25] Brazzle, Dokmeci, et al., describe this process:[26]

The mechanism of the etch is as follows. First, the XeF_2 adsorbs and dissociates to xenon (Xe) and fluorine (F) on the surface of silicon. Fluorine is the main etchant in the silicon etching process. The reaction describing the silicon with XeF_2 is

$$2\ XeF_2 + Si \rightarrow 2\ Xe + SiF_4$$

XeF_2 has a relatively high etch rate and does not require ion bombardment or external energy sources in order to etch silicon.

Safety:

Oxidizing, corrosive, toxic solid. May ignite or explode on contact with combustive material. Self-contained breathing apparatus and protective clothes must be worn by rescue worker.

References:

1. James L. Weeks, Max S. Matheson. "Xenon Difluoride". *Inorg. Synth.* 8. doi:10.1002/9780470132395.ch69.

2. Tius, M. A. (**1995**). "Xenon difluoride in synthesis". *Tetrahedron* **51** (24): 6605–6634.doi:10.1016/0040-4020(95)00362-C.

3. Chernick, CL and Claassen, HH and Fields, PR and Hyman, HH and Malm, JG and Manning, WM and Matheson, MS and Quarterman, LA and Schreiner, F. and Selig, HH and others (**1962**). "Fluorine Compounds of Xenon and Radon". *Science* 138 (3537): 136–138. Bibcode: 1962Sci..138.136C. doi: 10.1126/science.138.3537.136.PMID 17818399.

4. Hoppe, R. ; Daehne, W. ; Mattauch, H. ; Roedder, K. (**1962**). "Fluorination of Xenon".*Angew. Chem. Intern. Ed. Engl.* 1 (11): 599. doi:10.1002/anie.196205992.

5. Hoppe, R. (**1964**). "Die Valenzverbindungen der Edelgase". *Angewandte Chemie* 11: 455. doi:10.1002/ange.19640761103. First review on the subject by the pioneer of covalent noble gas compounds.

6. Weeks, J.; Matheson, M.; Chernick, C., (**1962**). "Photochemical Preparation of Xenon Difluoride" Photochemical Preparation of Xenon Difluoride". *J. Am. Chem. Soc.* 84 (23): 4612–4613. doi:10.1021/ja00882 a063.

7. Williamson, Stanley M.; Sladky, Friedrich O.; Bartlett, Neil (**1968**). "Xenon Difluoride".*Inorg. Synth.* 11: 147–151. doi:10.1002/9780470-132425.ch31.

8. Šmalc, Andrej; Lutar, Karel; Kinkead, Scott A. (**1992**). "Xenon Difluoride (Modification)".*Inorg. Synth.* 29: 1–4. doi:10.1002/9780470-132609.ch1.

9. Morrow, S. I.; Young, A. R. (**1965**). "The Reaction of Xenon with Dioxygen Difluoride. A New Method for the Synthesis of Xenon Difluoride". *Inorganic Chemistry* 4 (5): 759–760.doi:10.1021/ic50027 a038.

10. Harding, Charlie; Johnson, David Arthur; Janes, Rob (**2002**). *Elements of the* p *block*. Contributor Charlie Harding, David Arthur Johnson, Rob Janes. Royal Society of Chemistry (Great Britain), Open University. ISBN 0-85404-690-9.

11. Brown, D. R.; Clegg, M. J.; Downs, A. J.; Fowler, R. C.; Minihan, A. R.; Norris, J. R.; Stein, L. . (**1992**). "The dixenon(1+) cation: formation in the condensed phases and characterization by ESR, UV-visible, and Raman spectroscopy". *Inorganic Chemistry* 31(24): 5041–5052. doi:10.1021/ic 00050a023.

12. Stein, L. .; Henderson, W. W. (**1980**). "Production of dixenon cation by reversible oxidation of xenon".*Journal of the American Chemical Society* 102 (8): 2856-2857.doi:10.1021/ja00528a065.

13. Mackay, Kenneth Malcolm; Mackay, Rosemary Ann; Henderson, W. (**2002**). *Introduction to modern inorganic chemistry* (6th ed.). CRC Press. ISBN 0-7487-6420-8.

14. Egon Wiberg; Nils Wiberg; Arnold Frederick Holleman (**2001**). *Inorganic chemistry*. Academic Press. p. 422. ISBN 0-12-352651-5.

15. Melita Tramšek; Boris Žemva (**2006**). "Synthesis, Properties and Chemistry of Xenon(II) Fluoride". *Acta Chim. Slov.* 53 (2): 105–116. doi: 10.1002/chin. 200721209.

16. Tramšek, M.; Benkič, P.; Žemva, B. (**2004**). "First Compounds of Magnesium with XeF_2". *Inorg. Chem.* 43 (2): 699-703. doi:10.1021/ic 034826o.

17. Tramšek, M.; Benkič, P.; Žemva, B. (**2004**). "The First Compound Containing a Metal Center in a Homoleptic Environment of XeF_2 Molecules". *Angewandte Chemie International Edition* 43 (26): 3456. doi:10.1002/anie.200453802

18. Kim, M.; Debessai, M.; Yoo, C. S. (**2010**). "Two- and three-dimensional extended solids and metallization of compressed XeF2". *Nature Chemistry* 2 (9): 784–788. Bibcode:2010NatCh...2..784K. doi:10.1038/ nchem. 724.PMID 20729901.

19. Kurzydłowski, D.; Zaleski-Ejgierd, P.; Grochala, W.; Hoffmann, R. (**2011**). "Freezing in Resonance Structures for Better Packing: XeF_2 Becomes (XeF+)(F−) at Large Compression". *Inorganic Chemistry* 50 (8): 3832–3840. doi:10.1021/ic200371a. PMID 21438503.

20. W. Henderson (**2000**). *Main group chemistry*. Great Britain: Royal Society of Chemistry. p. 150. ISBN 0-85404-617-8.

21. Patrick, Timothy B.; Johri, Kamalesh K.; White, David H.; Bertrand, William S.; Mokhtar, Rodziah; Kilbourn, Michael R.; Welch, Michael J. (**1986**). "Replacement of the carboxylic acid function with fluorine". *Can. J. Chem.* 64: 138. doi:10.1139/v86-024.

22. Grakauskas, Vytautas (**1969**). "Aqueous fluorination of carboxylic acid salts". *J. Org. Chem.* 34 (8): 2446. doi:10.1021/jo01260a040.

23. Tamura Masanori; Takagi Toshiyuki; Shibakami Motonari; Quan Heng-Dao; Sekiya Akira (**1998**). "Fluorination of olefins with xenon difluoride-silicon

tetrafluoride". *Fusso Kagaku Toronkai Koen Yoshishu* (in Japanese) (Japan) 22: 62–63. Journal code: F0135B; accession code: 99A0711841.

24. Chang, F.; Yeh, R.; Lin, G.; Chu, P.;Hoffman, E.; Kruglick, E.; Pister, K.; Hecht, M;"Gas-phase silicon micromachining with xenon difluoride", SPIE Proc. V2641, **1995**, p. 117-128.

25. Chu, P.; Chen, J.; Chu, P.; Lin, G.; Huang, J.; Warneke, B; Pister, K.; "Controlled Pulse-Etching with Xenon Difluoride", Int. Conf. Solid State Sensors and Actuators, **1997** (Transducers 97), p. 665-668.

26. Brazzle, J.D.; Dokmeci, M.R.; Mastrangelo, C.H.; Modeling and characterization of sacrificial polysilicon etching using vapor-phase xenon difluoride, 17th IEEE International Conference on Micro Electro Mechanical Systems (MEMS), **2004**, p. 737-740.

3.1.41. Xenon hexafluoride

Xenon hexafluoride is a noble gas compound with the formula XeF_6 and the highest of the three known binary fluorides of xenon, the other two being XeF_2 and XeF_4. All known are exergonic and stable at normal temperatures. XeF_6 is the strongest fluorinating agent of the series. At room temperature, it is a colourless solid that readily sublimes into intensely yellow vapours.

Structure:	
IUPAC Name:	Xenone hexafluoride
Other Name:	Xenone-fluoride
CAS Number:	13693-09-9
Molecular formula:	XeF_6
Molar mass:	245.28 g/mol

Appearance:	-
Density:	3.56 g/cm^3
Melting point:	$49.25 \text{ °C} (120.65 \text{ °F}; 322.40 \text{ K})$
Boiling point:	$75.6 \text{ °C} (168.1 \text{ °F}; 348.8 \text{ K})$
Solubility in water:	Reacts

Preparation and properties:

Xenon hexafluoride can be prepared by long-term heating of XeF_2 at about 300 °C and pressure 6 MPa (60 atmospheres).

With NiF_2 as catalyst, however, this reaction can proceed at 120 °C even in xenon-fluorine molar ratios as low as 1:5.[1]

Hydrolysis

Xenon hexafluoride hydrolyzes stepwise, ultimately affording xenon trioxide:[2]

$$XeF_6 + H_2O \rightarrow XeOF_4 + 2 HF$$

$$XeOF_4 + H_2O \rightarrow XeO_2F_2 + 2 HF$$

$$XeO_2F_2 + H_2O \rightarrow XeO_3 + 2 HF$$

XeF_6 serves as a Lewis acid, binding one and two fluoride anions:

$$XeF_6 + F^- \rightarrow XeF_7^-$$

$$XeF_7^- + F^- \rightarrow XeF_2^-$$

With Fluoride Acceptor

XeF_6 reacts with strong fluoride acceptors such as RuF_5[3] and $BrF_3 \cdot AuF_3$[4] to form the XeF_5^+ cation:

$$XeF_6 + RuF_5 \rightarrow XeF_5^+RuF_6^-$$

$$XeF_6 + BrF_3 \cdot AuF_3 \rightarrow XeF+5AuF_4^- + BrF_3$$

Structure:

The structure of XeF_6 required several years to establish in contrast to the cases of XeF_2 and XeF_4. In the gas phase the compound is monomeric. VSEPR theory predicts that due to the presence of six fluoride ligands and one lone pair of electrons the structure lacks perfect octahedral symmetry, and indeed electron diffraction combined with high-level calculations indicate that the compound's point group is C_{3v}. The calculated energy for the point group O_h is only insignificantly higher, indicating that the minimum on the energy surface is very shallow. Konrad Seppelt, an authority on noble gas and fluorine chemistry, says, "The structure is best described in terms of a mobile electron pair that moves over the faces and edges of the octahedron and thus distorts it in a dynamic manner."[5]

^{129}Xe and ^{19}F NMR spectroscopy indicates that in solution the compound assumes a tetrameric structure: four equivalent xenon atoms are arranged in a tetrahedron surrounded by a fluctuating array of 24 fluorine atoms that interchange positions in a "cogwheel mechanism".

XeF_6 crystallizes in 6 possible modifications,[6] including one that contains XeF_5^+ ions with bridging F^- ions.[3]

Octafluoroxenates

Salts of the octafluoroxenate(VI) anion ($XeF2-8$) are very stable, decomposing only above 400 °C.[7][8][9] This anion has been shown to have square antiprismatic geometry, based on single-crystal X-ray counter analysis of its nitrosonium salt, $(NO)_2XeF_8$.[10] The sodium and potassium salts are formed directly from sodium fluoride and potassium fluoride:[9]

$2 NaF + XeF_6 \rightarrow Na_2XeF_8$

$2 KF + XeF_6 \rightarrow K_2XeF_8$

These are thermally less stable than the caesium and rubidium salts, which are synthesized by first forming the heptafluoroxenate salts:

$CsF + XeF_6 \rightarrow CsXeF_7$

$RbF + XeF_6 \rightarrow RbXeF_7$

which are then pyrolysed at 50 °C and 20 °C, respectively, to form the yellow[11] octafluoroxenate salts:[7][8][9]

$$2\,CsXeF_7 \rightarrow Cs_2XeF_8 + XeF_6$$

$$2\,RbXeF_7 \rightarrow Rb_2XeF_8 + XeF_6$$

These salts are hydrolysed by water, yielding various products containing xenon and oxygen.[9]

The two other binary fluorides of xenon do not form such stable adducts with fluoride.

Application:

Fluorination using Xenone hexafluoride.[12]

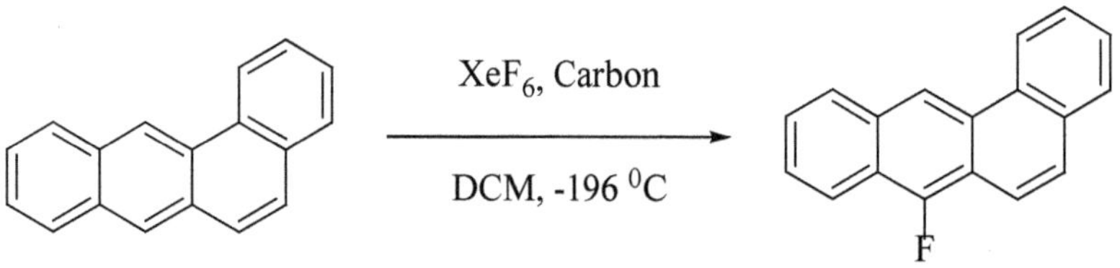

Safety:

Material is corrosive to the tissue of mucous membrane and upper respiratory tract. Do not breathing immediately evacuate all persons from danger area. Use proper protective equipment like mask, gloves, eye protection etc.

References:

1. Melita Tramšek; Boris Žemva (December 5, **2006**). "Synthesis, Properties and Chemistry of Xenon(II) Fluoride" (PDF). *Acta Chim. Slov.* 53 (2): 105–116. doi:10.1002/chin.200721209.

2. Appelman, E. H.; J. G. Malm (June **1964**). "Hydrolysis of Xenon Hexafluoride and the Aqueous Solution Chemistry of Xenon". *Journal of the American Chemical Society* 86 (11): 2141–2148.doi:10.1021/ja01065 a009.

3. James E. House (**2008**). *Inorganic Chemistry*. Academic Press. p. 569. ISBN 0-12-356786-6.

4. Cotton (**2007**). *Advanced Inorganic Chemistry* (6th ed.). Wiley-India. p. 591. ISBN 81-265-1338-1.

5. Seppelt, Konrad (June **1979**). "Recent Developments in the Chemistry of Some Electronegative Elements". *Accounts of Chemical Research* 12 (6): 211–216. doi:10.1021/ar50138a004.

6. Hoyer, S.; Emmler, K.; Seppelt, T. (October **2006**). "The structure of xenon hexafluoride in the solid state". *Journal of Fluorine Chemistry* 127 (10): 1415–1422. doi:10.1016 /j.jfluchem. 2006.04.014. ISSN 0022-1139.

7. Holleman, A. F.; Wiberg,, E. (**2001**). *Inorganic Chemistry*. San Diego: Academic Press. ISBN 0-12-352651-5.

8. Riedel, Erwin; Janiak, Christoph (**2007**). *Anorganische Chemie* (7th ed.). Walter de Gruyter. p. 393. ISBN 3-11-018903-8.

9. Chandra, Sulekh (**2004**). *Comprehensive Inorganic Chemistry*. New Age International. p. 308. ISBN 81-224-1512-1.

10. Peterson, W.; Holloway, H.; Coyle, A.; Williams, M. (Sep **1971**). "Antiprismatic Coordination about Xenon: the Structure of Nitrosonium Octafluoroxenate(VI)". Science 173 (4003): 1238–1239. Bibcode: 1971 Sci... 173.1238P.doi:10.1126/science.173.4003.1238.ISSN 0036-8075.

11. "Xenon". *Encyclopaedia Britannica*. Encyclopaedia Britannica Inc. **1995**.

12. Agranat, I. Et al. Synthesis. **1977**, p. 267-268.

3.1.42. XtalFluor-E

Diethylaminodifluorosulfinium tetrafluoroborate (XtalFluor-E) and morpholino-difluorosulfinium tetrafluoroborate (XtalFluor-M) are crystalline fluorinating agents that are more easily handled and significantly more stable than Deoxo-Fluor, DAST, and their analogues. These reagents can be prepared in a safer and more cost-efficient manner by avoiding the laborious and hazardous distillation of dialkylaminosulfur trifluorides. Unlike DAST, Deoxo-Fluor, and Fluolead,

XtalFluor reagents do not generate highly corrosive free-HF and therefore can be used in standard borosilicate vessels.

Structure:	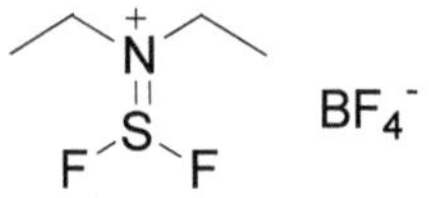
IUPAC Name:	*N*-(Difluoro-λ_4-sulfanylidene)-*N*-ethyl-ethanaminium tetrafluoroborate
Other Name:	*N,N*-Diethylamino-*S,S*-difluorosulfinium tetrafluoroborate, DAST difluorosulfinium salt
CAS Number:	63517-29-3
Molecular formula:	$C_4H_{10}BF_6NS$
Molar mass:	229.00 g/mol
Appearance:	Pale Brown powder
Density:	-
Melting point:	84-87 °C
Solubility in water:	Reacts
Solubility:	Oragnic solvents like DCM, MeCN

Preparation and properties:

Diethylaminotrimethylsilane react with the cold solution of Sulfur tetrafluoride in dichloromethane below −65 °C under nitrogen atmosphere to yield dark amber solution of diethylaminosulfur trifluoride. Resulting dark amber solution reacts with boron trifluoride tetrahydrofuran complex to give Xtalfluor-E.[1][2]

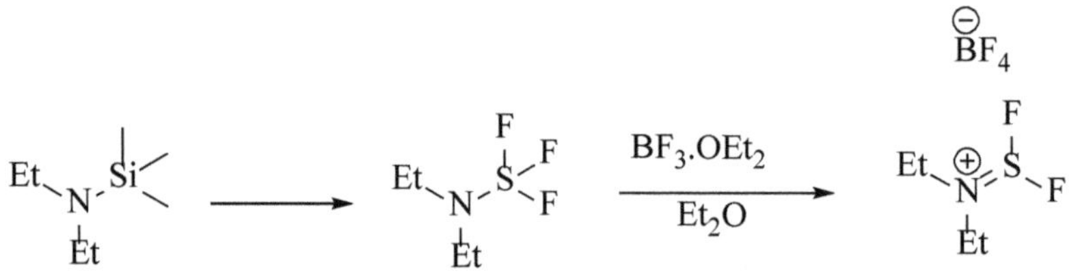

Scheme 1

Applications:

Some applications related to Xtalfluor-E are given below.

Fluorination of hydroxyls

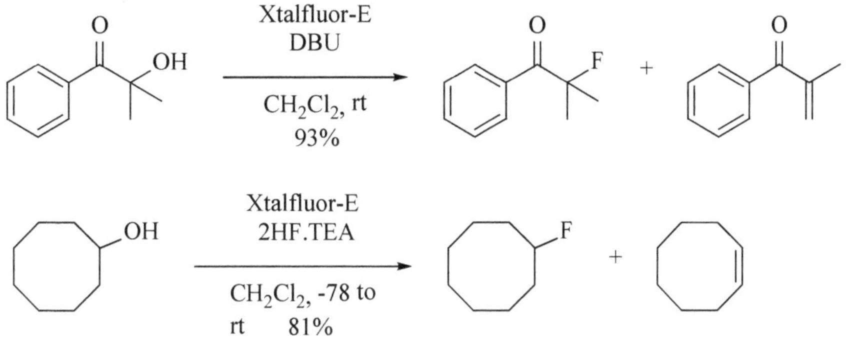

Scheme 2

Fluorination of carbonyls

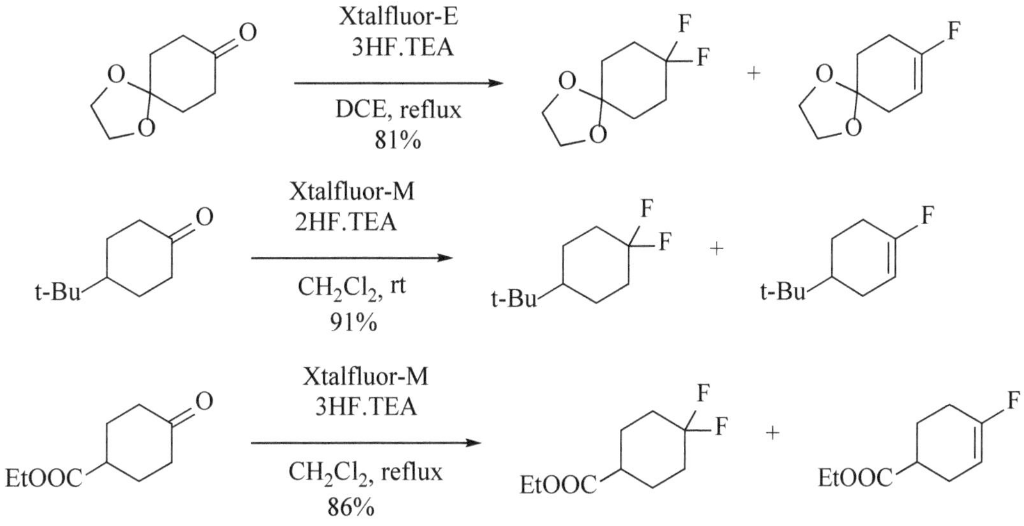

Scheme 3

Safety:

It is generate Hydrofluoric acid in the contac of moisture. Hydrofluoric (HF) acid burns require immediate and specialized first aid and medical treatment. Symptoms may be delayed up to 24 hours depending on the concentration of HF. After decontamination with water, further damage can occur due to penetration/absorption of the fluoride ion. Treatment should be directed toward binding the fluoride ion as well as the effects of exposure. Skin exposures can be treated with a 2.5% calcium gluconate gel repeated until burning ceases. More serious skin exposures may require subcutaneous calcium gluconate except for digital areas unless the physician is experienced in this technique, due to the potential for tissue injury from increased pressure. Absorption can readily occur through the subungual areas and should be considered when undergoing decontamination. Prevention of absorption of the fluoride ion in cases of ingestion can be obtained by giving milk, chewable calcium carbonate tablets or Milk of Magnesia to conscious victims. Conditions such as hypocalcaemia, hypomagnesaemia and cardiac arrhythmias should be monitored for, since they can occur after exposure. Consult a physician. Show this safety data sheet to the doctor in attendance.

References:

1. Couturier, M. Et. Al. *Org. Lett.* **2009**, 11, 5050.
2. Couturier, M. Et. Al. *J. Org. Chem.* **2010**, 75, 3401.

3.1.43. XtalFluor-M

Structure:	
IUPAC Name:	4-(Difluoro-λ^4-sulfanylidene)morpholin-4-ium tetrafluoroborate
Other Name:	Morpho-DAST difluorosulfinium salt, Xtalfluor-M
CAS Number:	63517-33-9
Molecular formula:	$C_4H_8BF_6NOS$
Molar mass:	242.98 g/mol

Appearance:	Light yellow liquid
Density:	1.436 g/cm^3 at 25 °C
Melting point:	117-126 °C
Boiling point:	-
Solubility in water:	Reacts violently
Solubility:	Oragnic solvents like DCM, MeCN

Preparation and properties:

Morpho-DAST react with BF$_3$-etherate to give Xtalfluor-M.[1]

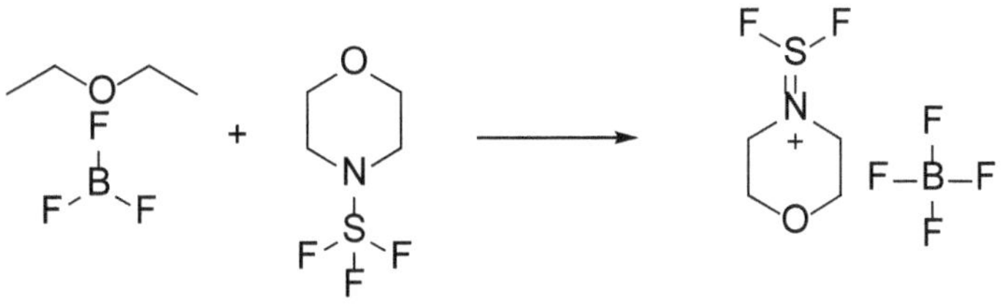

Scheme 1

Applications:

Some applications related to Xtalfluor-M are given below.[2][3][4]

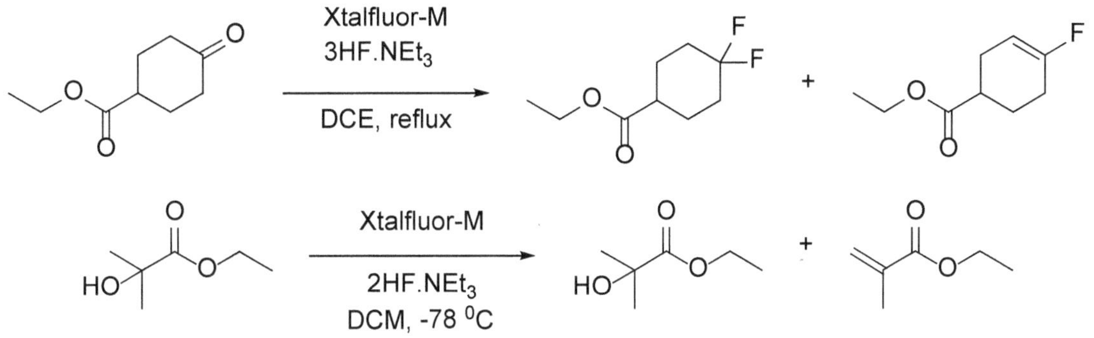

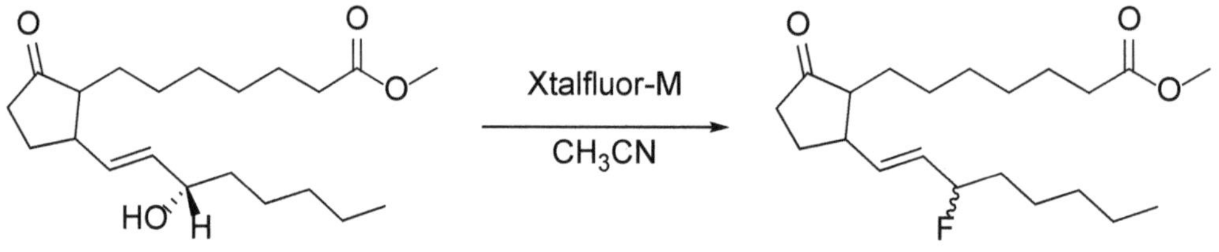

Scheme 2

Safety:

It is generate Hydrofluoric acid in the contact of moisture. Hydrofluoric (HF) acid burns require immediate and specialized first aid and medical treatment. Symptoms may be delayed up to 24 hours depending on the concentration of HF. After decontamination with water, further damage can occur due to penetration/absorption of the fluoride ion. Treatment should be directed toward binding the fluoride ion as well as the effects of exposure. Skin exposures can be treated with a 2.5% calcium gluconate gel repeated until burning ceases. More serious skin exposures may require subcutaneous calcium gluconate except for digital areas unless the physician is experienced in this technique, due to the potential for tissue injury from increased pressure. Absorption can readily occur through the subungual areas and should be considered when undergoing decontamination. Prevention of absorption of the fluoride ion in cases of ingestion can be obtained by giving milk, chewable calcium carbonate tablets or Milk of Magnesia to conscious victims. Conditions such as hypocalcaemia, hypomagnesaemia and cardiac arrhythmias should be monitored for, since they can occur after exposure. Consult a physician. Show this safety data sheet to the doctor in attendance.

References:

1. *Zhurnal Organicheskoi Khimi.* **1977,** Vol. 13, p. 1025-1026.
2. Couturier, M. Et. Al. *Org. Lett.* **2009,** Vol. 11, 5050.
3. Couturier, M. Et. Al. *J. Org. Chem.* **2010,** Vol. 75, 3401.
4. Bezuglov V. V.; Pashinik, V. E.; Tovstenko, V. I.; Markovskii, L. N.; Freimanis, Y. A.; Serkov, I. V. *Russ. J. Bioorg. Chem.* **1996,** Vol. 22, 688.

3.1.44. Yarovenko's reagent

Structure:

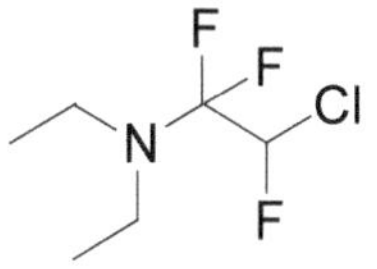

IUPAC Name: N,N-Diethyl-2-chloro-1,1,2-trifluoroethylamine

Other Name: Yarovenko's reagent, 2-Chloro-1,1,2-
                        trifluorotriethylamine

CAS Number: 357-83-5

Molecular formula: $C_6H_{11}ClF_3N$

Molar mass: 189.61 g/mol

Appearance: Liquid

Density: 1.19 g/cm^3

Melting point: -

Boiling point: 33-34/6 mm Hg

Solubility in water: Reacts

Solubility: Reacts with alcohol and amine

                        Soluble in ether, CH_2Cl_2, acetonitrile.

Preparation and properties:

N,N-Diethyl-2-chloro-1,1,2-trifluoroethylamine (Yarovenko's-reagent) synthe-size
by the reaction between diethylamine and 1-chloro-1,2,2-trifluoroethene.[1][2]

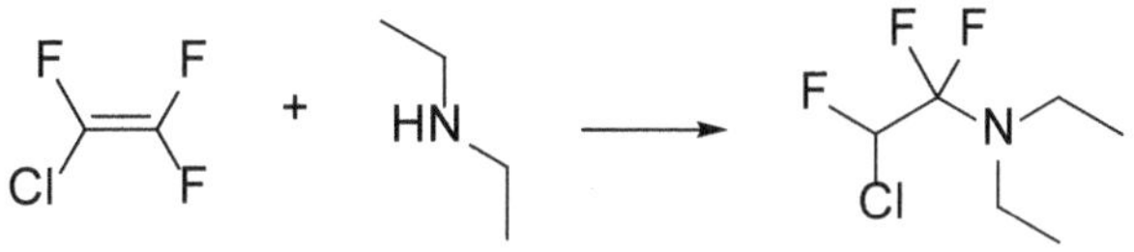

Scheme 1

Applications:

Converts alcohols to alkyl fluorides, carboxylic acids to acid fluorides; used as acylating agent. one of the example related to conversion of hydroxyl to fluoride.[3]

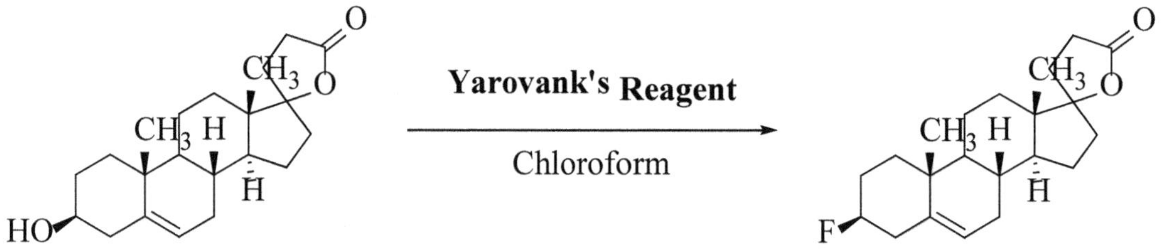

Scheme 2

Yarovanko's reagent also reacts with hydrogen cynide to give 2-(chlorofluoro-methyl)-2-(diethylamino)malononitrile.[4]

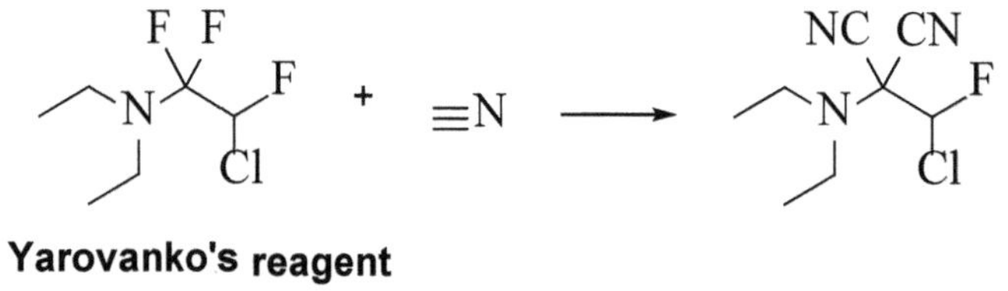

Yarovanko's reagent

Scheme 3

Safety:

Corrosive. In combustion emits toxic fumes. Carbon oxides. Nitrogen oxides (NOx). Hydrogen fluoride (HF). Hydrogen chloride (HCl). Wear self-contained breathing apparatus. Wear protective clothing to prevent contact with skin and eyes. Avoid direct contact with the substance. Ensure there is sufficient ventilation of the area. Do not handle in a confined space. Avoid the formation or spread of mists in the air. Only use in fume hood.

References:

1. *Journal of Chemical Research Miniprint.* **1998,** #1, p.301-326.

2. *Journal of American chemical society.* **1950,** vol. 172, p.3646-3647.

3. *Acta chimica Academiae scientiarum Hungaricae.* **1980,** Vol. 103, #2, p.231-240.

4. *Journal of American Chemical Society.* **1960,** Vol.82, p.5116-5122.

www.ingramcontent.com/pod-product-compliance
Lightning Source LLC
Chambersburg PA
CBHW050714180526
45159CB00003B/1026